Hybrid Rail Vehicles

Aleksandr Luvishis,
Lev Luvishis, Igor Luvishis

Front cover illustration by Maxim Kuperman.

ISBN 978-0-578-04577-1

Introduction **6**

1. What is a hybrid rail vehicle?. 7

1.1 General description. Modern hybrid rail vehicles 7

1.2 Hybrid rail vehicle - definition 9

1.3 Features of hybrid rail vehicles 12

1.4 Energy economy - hybrid rail vehicles with energy storage 14

    1.4.1 The braking energy is used for the next acceleration; the stops are

        frequent, short segments 14

    1.4.2. The braking energy is stored in the energy system; the stops are

        few, long segments 18

1.5 Energy and power of regenerative braking 20

    1.5.1 The hybrid rail vehicles with frequent stops 20

    1.5.2 Hybrid switcher locomotive 22

    1.5.3 Hybrid freight locomotive 22

2. General elements of hybrid rail vehicles **25**

2.1 Prime movers and motors 25

    2.1.1 The diesel engines 25

    2.1.2 Microturbines 25

    2.1.3 Fuel cells 26

    2.1.4 Main generators 29

    2.1.5 Traction motors 30

2.2 Converters. 30

    2.2.1. The components of the modern traction converters 30

    2.2.2 3-phase Rectifiers Bridge 31

    2.2.3. The choppers 31

    2.2.4. PWM inverter 32

2.3. Modern energy storage devices 33

    2.3.1. Battery Packs 33

        2.3.1.1 Lead Acid Battery 33

2.3.1.2. Nickel Metal Hydride (NiMH) Battery 34

2.3.1.3. ZEBRA (Sodium/Nickel – Chloride) Battery. 35

2.3.1.4. Lithium Ion Battery 36

2.3.1.5. Lithium Ion Polymer Battery 37

2.3.2. Ultracapacitors 37

2.3.3. Flywheels 38

3. Conventional and dual-mode rolling stock. **42**

3.1 Dual-mode tram-trains. 42

3.2 Diesel and dual-mode trains. 42

3.3 Diesel and dual-mode locomotives 48

4. DC traction units with on-board energy storage. **56**

4.1. Light Rail Vehicle (LRV) with on-board ultracapacitor energy storage

(Mannheim, Germany). 57

4.2. Tram Citadis with on-board battery energy storage (France) 57

4.3 Tram Citadis with on-board flywheel energy storage (France) 57

4.4. Tram with Li-Ion battery energy storage (Japan) 58

4.5. Electric Multiple-Unit Train with on-board Li-Ion battery energy storage

(Japan) 59

4.6. Railcar Parry People Mover PPM 35 (England) 60

5. Examples of the real hybrid trams, hybrid diesel trains

and hybrid locomotives. **62**

5.1. The world's first hybrid railcar and the hybrid diesel-battery

multiply unit (DMU) Kiha E200 (Japan) 62

5.2 Hybrid regional innovative train LIREX (Germany) 64

5.3 Hybrid railcars PPM 50 and PPM 80 (England) 67

5.4 Modernized Hybrid Switch Locomotives. 68

5.4.1 Hybrid Switch Locomotives Green Goat (GG20B) 68

5.4.2 Hybrid Switch Locomotive Green Kid (GK 10B) 71

5.4.3 Hybrid Road Switcher RP20BH 71

**6. Hybrid trams, trains and locomotives with energy storage –
latest developments** 73

6.1 Hybrid trams, trains and locomotives with energy storage and heat engines. . . . 73

    6.1.1. Hybrid tram with microturbine and flywheel energy storage
       (project) 73

    6.1.2 Hybrid diesel train with ultracapacitor energy storage
       (Mitrac Energy Saver, Bombardier Germany). 74

    6.1.3. British Green Goat switcher 75

    6.1.4 Swedish Green Goat switcher 76

    6.1.5 Hybrid modernized switcher locomotives of Alstom Transport
       (France) 76

    6.1.6 Hybrid turbine-electric switcher locomotive (Russia). 77

    6.1.7. GE hybrid diesel locomotive (USA) 77

    6.1.8 Hybrid High Speed Train (Great Britain) 78

6.2 Hybrid trams, trains and locomotives with fuel cell energy storage 79

    6.2.1 Hybrid fuel cell tram (FULLTRAM project) 79

    6.2.2 World's First Fuel Cell Hybrid Railcar (East Japan Railway) 80

    6.2.3 Fuel cell train project of Japan Railway
       Technical Research Institute (RTRI) 81

    6.2.4 Fuel cell-battery Hybrid Switcher 82

7. The analysis of the modern hybrid trains and hybrid switcher 85

7.1 The features of the hybrid diesel multiple unit 85

7.2. New features of hybrid switcher Green Goat 20, Green Kid 10B 86

7.3. A new perspective for hybrid switchers. 87

7.4 The features of the regenerative braking on hybrid switchers. 88

7.5 Modern energy storage systems. 90

8. Conclusions 97

Introduction

Reduction in fossil fuel consumption has always been a concern and an objective for railroads.

However, this reduction became of particular importance in recent years when it was found that reduction in diesel oil consumption is closely connected to the reduction in emissions and helps ecology improvement. Greenhouse gases, such as carbon dioxide and methane, are produced mainly through the burning of fossil fuels (including diesel oil) and are contributing to a potentially disastrous change in the global climate.

Kyoto Commitment asserts that the task of ecological improvement is the most important task for humanity. According to the Kyoto Commitment, the amount of the exhaust emissions (per year) in the future has to be maintained at the 1990 level.

Combined exhaust emissions quantity produced by the major industrialized countries is 55% of the world's quantity. The volume of exhaust emissions produced by locomotives and diesel trains, in total volume of all exhaust emissions, varies from country to country. In every country railroads produce much less exhaust emissions than motor transportation.

Let's assume that all of the exhaust emissions of carbon dioxide (from all kinds of transport) in the USA were equal to 100% in 2001. Then the exhaust emissions from motor transport were 78.4%, while they were only 1.9% from locomotives. It means that the exhaust emissions of locomotives are equal to 2.42% of the exhaust emissions of motor transport. However, the task to decrease the consumption of fossil fuel on the railways of the industrialized countries of the world and railways of the USA, in particular, remains significant.

Many countries around the world have hybrid cars and hybrid buses on their roads today. More and more countries are just starting to use hybrids. The major reason for the new interest in hybrids is that the hybrid motor transportation considerably reduces consumption of fossil fuels and, therefore, reduces the exhaust emissions. Hybrid cars and hybrid buses are well studied today. There is a great amount of literature covering modern hybrid cars. At the same time, there is not enough literature on the market today that describes operation and maintenance as well as different features of the hybrid rail vehicles. This book is an attempt to make up for this deficiency.

1. WHAT IS A HYBRID RAIL VEHICLE?

1.1. General description. Modern hybrid rail vehicles

Traditionally rail vehicles are divided into two main categories: electric rail vehicles and rail vehicles with some form of the heat engine(usually diesel).

Rail vehicles with heat engines today include railcars, diesel trains and diesel locomotives. The majority of rail vehicles with heat engines also have electric drive.

The heat engine consumes diesel fuel and is the main source of CO_2 and other pollution gases. In many cases the diesel engine of a rail vehicle is inefficient.

In contrast to electric train and electric locomotive the diesel train and diesel locomotive were not originally designed to be able to recover the braking energy. Braking energy of the diesel trains and diesel locomotives is dissipated into conventional mechanical brakes as well as braking resistors if the vehicle is equipped with dynamic braking system. Inability to recover breaking energy adds to the inefficiency of this category of rail vehicles.

Hybrid rail vehicles reduce the fuel consumption and therefore also reduce CO_2 and other pollutant emissions. Hybrid rail vehicles typically use on-board energy storages systems. On-board energy storage system allows the reduction in power of the heat engine. The heat engine drive is optimized and its efficiency increases. The energy storage system provides or absorbs the difference between the power provided by the heat engine and that one needed for the drives of vehicle or train or locomotive.

The energy storage system is charged by the energy recovered during the mode of regenerative braking. The stored energy can be reused during the future acceleration or can be used for example for feeding of vehicle's auxiliary system circuits.

The battery packs, flywheels, and ultracapacitors are used as components of modern energy storage systems. These devices are covered in next chapter.

The main power circuit of the modern rail hybrid vehicle with heat engine is shown in **Figure 1.1.**

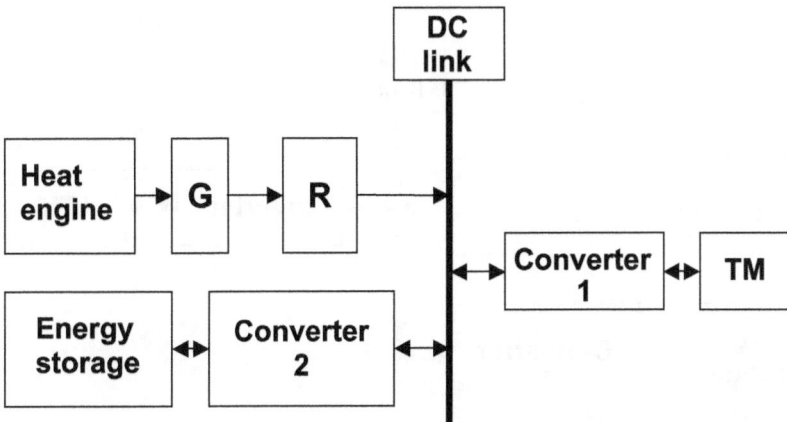

Figure 1.1   **Main circuits of the hybrid rail vehicle with heat engine and AC drive.**

The main circuit of the modern hybrid vehicle has traction drive and energy storage. The heat engine turns the main generator (G). The heat engine can be a diesel engine or a gas turbine. The voltage of the generator is rectified by a 3-phase diode bridge (R). The output of the diode bridge is a DC link. The energy storage connects to the DC link. In general case there is a converter (Converter 2) between the energy storage and the DC link. The type of this converter varies depending upon energy storage devices. Sometimes the output of the battery bank that is used as energy storage connects with the DC link by a converter, sometimes the output of the battery bank is connected directly with the DC link without a converter. If the flywheel is used as energy storage the inverter with pulse width modulation (PWM) is used as Converter 2, In case the ultra-capacitors are used as energy storage devices a chopper is used as Converter 2.

The converter 1 is fed from the DC link and feeds the traction motors (TM). If the vehicle has the DC drive the chopper is used as Converter 1 and TM are DC traction motors. If the vehicle has the asynchronous traction drive the inverter with pulse width modulation (PWM) is used as Converter 1 and TM are asynchronous traction motors.

The advantages of the hybrid rail vehicle with heat engine are as follows:
• Reduction of fuel consumption by recuperating braking energy
• Optimal, effective design of the heat engine
• Increased power during acceleration (booster regime)
• No new infrastructure is required
• Noise reduction during starting in the station
• Emission-free operation on short sections of the line (tunnels, stations)
• Auxiliary power supply when idling with the heat engine stopped

The main power circuit of the modern rail hybrid vehicle with fuel cells and asynchronous traction drive is shown in **Figure 1.2**. The Converter 1 is fed by the fuel cells and passes the energy to the DC link. Energy storage system is identical to the one used in hybrids with heat engine.

The inverter with pulse width modulation (PWM) is fed from the DC link and feeds the asynchronous traction motor (ATM).

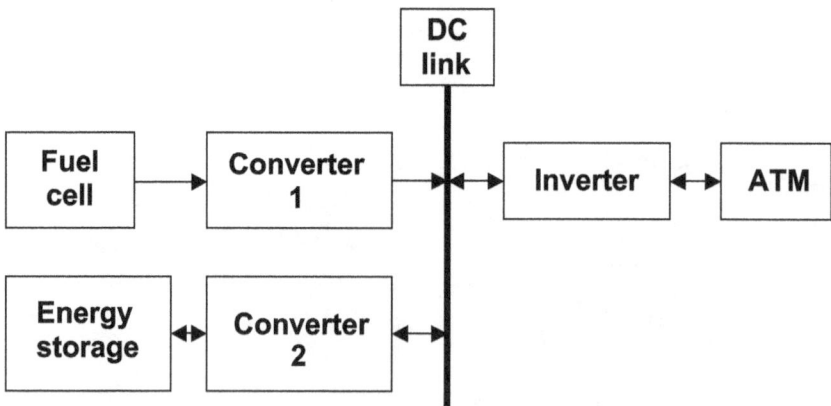

Figure 1.2. **Main circuits of the hybrid rail vehicle with fuel cells and AC drive**

The energy flow diagrams of the hybrid diesel system are shown in **Figures 1.3 -1.5.** The energy flows from the diesel-generator set as well as from the energy storage system (as long as it is possible) to the traction motor in the traction mode. The direction of the energy flow reverses in the regenerative mode. In this mode the energy flows from the traction motor to the energy storage and is used for additional charging. Figure 1.5 shows that the energy storage system could be charged by the traction motor (in the regenerative mode) and by the diesel-generator set.

1.2. Hybrid rail vehicle-definition

When different people use the term hybrid rail vehicle they mean different things. The following definition was used widely in the past: "The hybrid vehicle is the vehicle

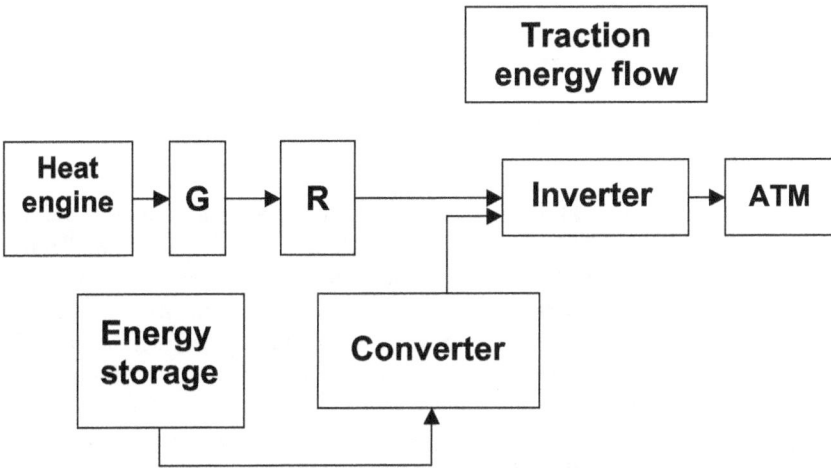

Figure 1.3   Energy flow – traction mode

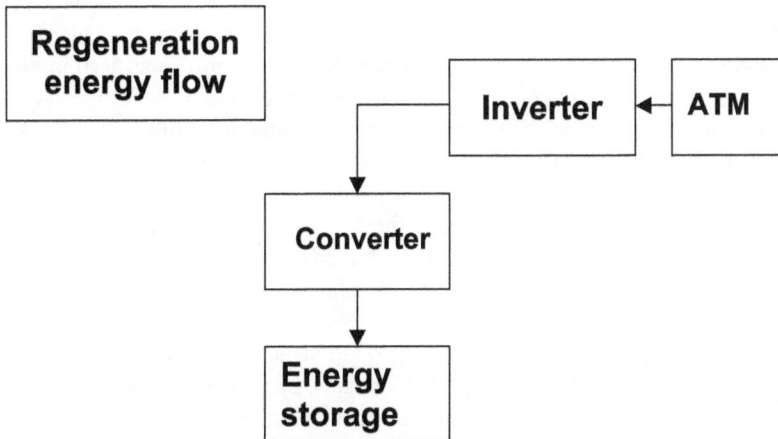

Figure 1.4 Energy flow – regenerative mode

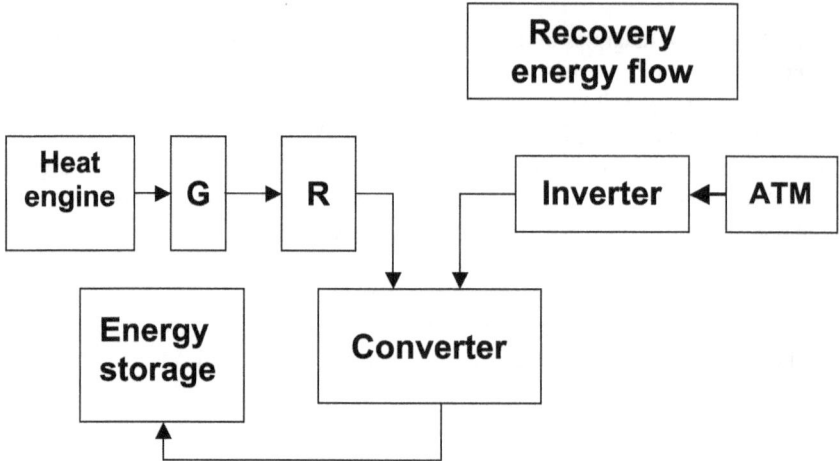

Figure 1.5 Energy flow – recovery mode

that has two or more sources of power". This definition is not applicable to hybrid rail vehicles. In 2003, the Union Nation (UN) defined a "hybrid vehicle" as follows:" A hybrid vehicle is a vehicle with at least two different energy converters and two energy storage systems (on-board the vehicle) for the purpose of vehicle propulsion." [1.1]

Let's look at a hybrid synergy drive installed in an automobile. Such automobile is equipped with both gas/petrol engine and an electric motor as two different energy converters. It also has a gas tank and a battery bank used as storage systems, therefore satisfying the UN definition. Hybrid synergy drive is a set of the hybrid car technologies developed by Toyota and used in the company's car models: Prius, Highlander Hybrid, Camry Hybrid, Lexus RX 400h and Lexus GS 450h. It combines an electric drive, internal combustion engine (ICE) and an electro-mechanical transmission.

When we use a term "hybrid car" today, we are referring to the cars with a hybrid synergy drive. However, the original term "hybrid car" had a different meaning. Prior to its modern meaning of hybrid propulsion, the word "hybrid" was used in the United States to mean a vehicle of mixed national origin; generally, a European car fitted with American mechanical components. This meaning is no longer widely known. Some sources had have also referred to flexible-fuel vehicles as "hybrid", because those vehicles can use a mixture of different fuels–usually gasoline and ethanol alcohol. The flexible-fuel vehicles are not "hybrid" vehicles in its modern meaning. They do not satisfy the UN definition, which is used today.

Let's now explore the term "hybrid rail vehicle". We think that a hybrid rail vehicle today is a rail vehicle that satisfies the UN definition.

Main power circuits of the modern hybrid rail vehicles are shown in Figure 1.1 and Figure 1.2. The modern hybrid diesel rail vehicle carries a heat engine and an electric motor as energy converters. It also has a diesel fuel tank and a battery bank representing the storage systems, therefore satisfying the UN definition. The modern hybrid fuel cells rail vehicle carries fuel cells and an electric motor as energy converters. It is also

```
                  ┌─────────────────────────────┐
                  │      Rolling stock with the │
                  │      energy storage system  │
                  └─────────────────────────────┘
                     │                      │
              ┌──────────────┐       ┌──────────────────┐
              │Hybrid vehicles│      │Electric vehicles │
              └──────────────┘       └──────────────────┘
               │          │            │            │
        ┌─────────┐ ┌─────────┐  ┌─────────┐  ┌──────────┐
        │ Diesel  │ │ Diesel  │  │ Subway  │  │ Electric │
        │locomotives│ │switchers│  │ trains  │  │locomotives│
        └─────────┘ └─────────┘  └─────────┘  └──────────┘
          │          │             │             │
      ┌─────────┐ ┌─────────┐  ┌─────────┐  ┌──────────┐
      │Fuel cell│ │  DEMUs  │  │  EMUs   │  │ Trams&   │
      │vehicles │ │         │  │         │  │  LRV     │
      └─────────┘ └─────────┘  └─────────┘  └──────────┘
          │
    ┌──────────────┐
    │ Tram-trains  │
    │with heat engine│
    └──────────────┘
```

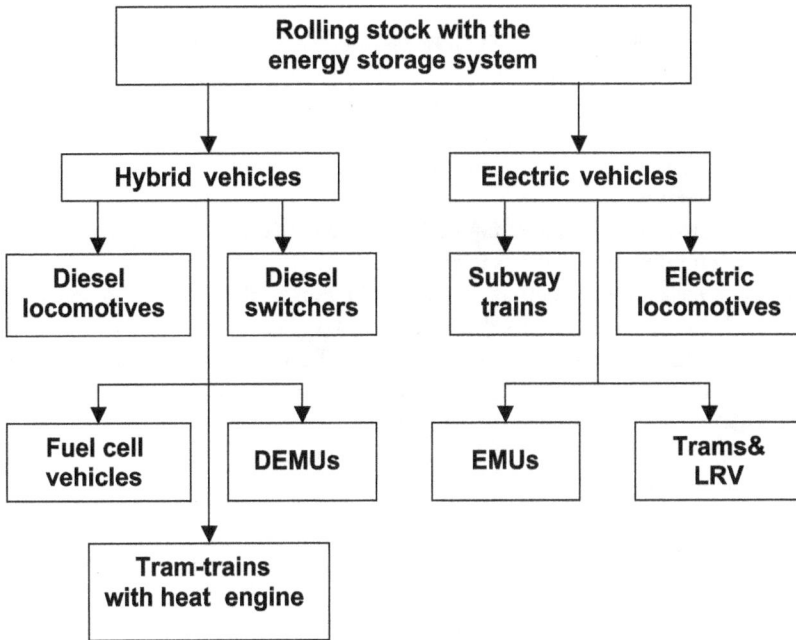

Figure 1.6 Rolling stock with the energy storage system

equipped with a hydrogen fuel tank and a battery bank representing the storage systems, therefore also satisfying the UN definition.

Just not long ago, in Germany, the dual-mode rail vehicles were referred to as "hybrid rail vehicles". This definition still exists today in some literature and is causing a lot of confusion. Modern dual-mode tram trains are called "hybrid trains" pretty often too.

However dual-mode vehicles do not satisfy UN "hybrid vehicle" definition. A dual-mode rail vehicle is a vehicle that can run on power from two different sources, typically electricity from overhead lines or a ground level power supply or alternative from an internal combustion engine burning liquid fuels or gas. Dual mode vehicles on contrary do not have two storage systems on-board the vehicle and, therefore, do not satisfy the UN definition. This is why we won't consider the dual-mode rail vehicles to be hybrid rail vehicles in this work.

At the same time not every vehicle that has its electric storage on-board could be referred to as hybrid rail vehicle. The Japanese scientists use this term when describing a vehicle, that is fed by a contact feeder line (trolley) and an on-board electric energy storage device. However, this vehicle has only one energy converter and one storage system on board and, therefore, does not satisfy the UN definition. This is just a DC traction unit with on-board energy storage.

The examples of the modern hybrid rail vehicles, dual mode rail vehicles and DC traction units with energy storage systems are described in the following chapters of this work.

Energy storage system is essential to the modern rolling stock. It provides the ability to store the kinetic energy during the brake time and feed it back during acceleration.

The energy storage system can be mounted on diesel rail vehicles as well as electric rail vehicles. The electric rail vehicles with energy storage system have advantages in comparison to electric rail vehicles without it for the following reasons:
- Increased reliability in the regeneration mode
- The increase in value of using regenerative energy
- Energy economy
- The reduction in peak power required from the network
- The reduction in voltage drops via contact wires
- Ability to pass through sections of a limited length without overhead wires.

Today all rail vehicles can be divided into two classes: conventional vehicles and vehicles with energy storage systems. Also, the vehicles with energy storage system can be divided into two classes as well: hybrid vehicles and electric vehicles **(Figure 1.6).**

Hybrid rail vehicles with the energy storage system include:
- diesel locomotives,
- diesel trains,
- diesel switchers,
- tram-trains with heat engines,
- fuel cell vehicles.

DC traction units with the energy storage system include:
- electric locomotives
- EMUs
- subway trains
- trams
- LRV vehicles.

1.3 Features of hybrid rail vehicles

The Toyota Prius is the world's first commercially mass-produced and marketed hybrid automobile. It went on sale in Japan in 1997, and worldwide in 2001. By the end of 2003, nearly 160,000 units had been produced for sale in Japan, Europe, and North America. Hybrid buses became popular in recent years as well. 13000 hybrid buses, manufactured by General Motors, are daily used in nine different cities across the USA. The hybrid cars and buses have good future prospects and will become even more popular in the years to come.

Our comparison shows that hybrid autos and buses have both similarities and differences with hybrid rail vehicles. The similarities include:
- reduced of energy consumption, the fuel economy,
- reduced of emissions,
- providing the ability to store the kinetic energy during the brake time and feed it back during acceleration
- using heat engine in an optimal mode
- design and implementation of the common components: electric traction motors, control systems, generators and converters

The differences include:
- Rolling resistance

The rolling resistance of the rail hybrid vehicles is less than the rolling resistance of hybrid road vehicles. The rolling resistance of the tram, diesel train or diesel locomotive is about 6-8 times lesser than the rolling resistance of the buses. The rolling resistance coefficient for train's steel on steel equals 0.001-0.0025. The rolling resistance coefficient for ordinary car's tires on concrete equals 0.01-0.015. The small rolling resistance of the rail vehicles carries a double impact. During bus acceleration additional power loss is observed; during bus braking the amount of energy that can be recuperated decreases.

• Common components

Energy storage systems could be easily implemented on most modern diesel locomotives and diesel trains. They have electric drives, electric motors, generators, and converters. The repairmen know the features of this equipment. The rail vehicles with asynchronous traction drive are especially close to hybrid rail vehicles. Common autos and buses do not have electric drive. The electric motors, generators, and converters are new for the road vehicle design.

• Size and weight

The rail hybrid vehicles require energy storage systems which size and weight are much greater compared with the road hybrid vehicles.

Modern energy storage devices store the braking energy on-board for use in the subsequent acceleration phase. This creates the possibility of an effective brake energy recovery in diesel-electric vehicles. With battery boost, the diesel engine can be made smaller and lighter and the train can carry less fuel. Generally, the hybrid vehicle is heavier than the standard vehicle of the same class due to the added weight of the energy storage system (ESS). However, in some cases, with ESS addition, the weight of the diesel engine and the cooling system can be reduced. That would lead to the weight reduction of the hybrid rail vehicle.

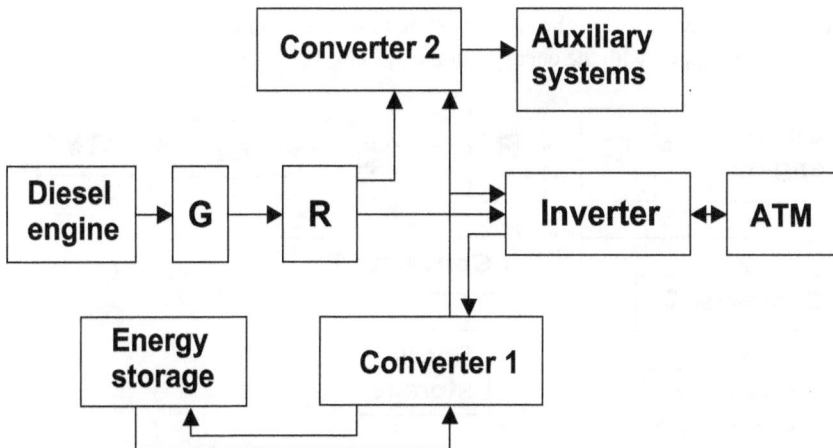

Figure 1.7 First variant of auxiliary system circuit

Adequate supply of power to the auxiliary services guarantees normal operation of the rail traction vehicle. The auxiliary systems of a diesel-electric vehicle include:
- the brake system compressors
- a fuel pump
- a lubrication fuel pump
- a fan of the diesel water radiator
- ventilators of traction motors and converters.

The auxiliary systems of the past were fed directly by a mechanical link with the diesel engine. The auxiliary systems of the diesel-electric vehicles with asynchronous traction drive are fed by the DC link, which connects to the rectifier in the output of the diesel-generator set. In case the vehicle has electric braking (regenerative or dynamic), part of the braking energy will be used to feed the auxiliary systems. Rail hybrid vehicles with asynchronous traction drive also use the feeding of the auxiliary systems from the DC link. In this case the auxiliary systems are fed not only from the diesel-generator set but from the energy storage system as well **(Figure 1.7)**. It is possible that in the future the auxiliary systems will be fed directly and only from the energy storage systems regardless of operation mode. **(Figure 1.8).**

1.4. Energy economy–hybrid rail vehicles with energy storage.
1.4.1 The braking energy is used for the next acceleration; the stops are frequent, short segments.

Hybrid rail vehicles, which are currently used or planned to be used in city transportation (fuel cell trams, fuel cell subway cars, fuel cells tram-trains) and in suburban transportation (diesel-electric multiple units, diesel-electric tram-trains, fuel cell multiple units), make frequent stops. The majority of switcher locomotives make frequent stops as well.

Let's compare the amount of energy needed to be produced by a diesel engine for a traditional diesel-electric vehicle as well as hybrid diesel-electric vehicle **(Figure 1.1)**. Both vehicles have the asynchronous motors, though the traditional diesel-electric vehicle is not equipped with the energy storage system.

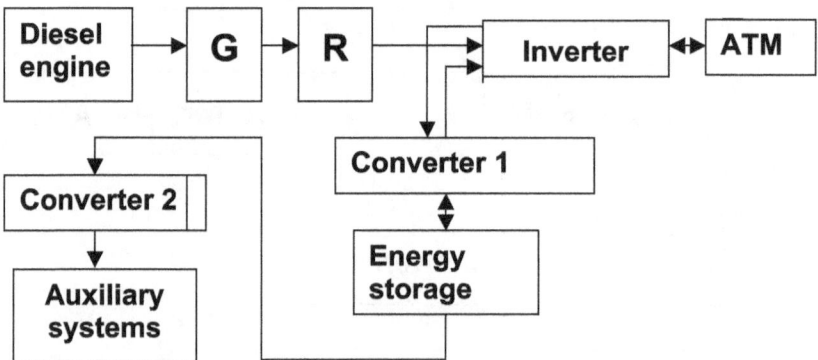

Figure 1.8 Second variant of auxiliary system circuit

For us to be able to get an estimate in energy consumption reduction in the case where energy storage system is utilized, let's suppose the following:

• The vehicle passes a road section that consists of "n" segments ("n" stations). All segments lengths are equal.
• The mechanical energy in the output shaft of the traction motor or motors A mech is the same for every section and also the same for vehicles with or without energy storage system.
• The energy storage system has minimal charge before first braking.

The mechanical energy in the output shaft of the traction motor or motors A mech is equal to:

$$A_{mech} = A_{dl} * \eta_{gen}, \text{ where:}$$

A_{dl} – diesel generated energy (diesel engine–main generator–rectifier) in the DC link input.

$$\eta_{gen} \text{ - total efficiency,} \tag{1}$$
$$\eta_{gen} = \eta_i * \eta_m;$$
η_i – the efficiency of the inverter, η_m – the efficiency of the traction motor

On the other hand: $A_{mech} = A_{mr} + A_b$, where: $\tag{2}$

A_{mr} - the work of train resistance forces (the sum of the rolling resistance, grade resistance and curve resistance),

$$A_b = A_k - \Delta A_{mr}, \text{ where:} \tag{3}$$

A_k - the kinetic energy of the vehicle before the braking, ΔA_{mr} – the work of train resistance forces in the braking mode.

In mode of regenerative braking
$A_b = E_{RB}$, where E_{RB} – the generated energy during regenerative braking or energy of regenerative braking (EORB)

From (2) and (3) follows:

$$A_{mech} = A_k + A_{mr} - \Delta A_{mr} \tag{4}$$

The energy AES, that charges the energy storage system during the regenerative regime equals:

$$A_{ES} = (A_k - \Delta A_{mr}) * \eta_{ml} * \eta_i * \eta_c * \eta_{ch}; \tag{5}$$

where: η_{ml} – the efficiency of the traction motor operating in the generator mode, η_c – the efficiency of the energy storage system's converter, η_{ch} – the efficiency of the energy storage system, operating in the charging mode.

In mode of regenerative braking we can wrote the equation (5) as:

$$A_{ES} = E_{RB} * \eta_{m1} * \eta_i * \eta_c * \eta_{ch};$$

If we designate $\eta_{ES} = \eta_{m1} * \eta_i * \eta_{c^2} * \eta_{ch} * \eta_{rech}$, then we can modify the equation to:

$$A_{ES} = E_{RB} * \eta_{ES} \tag{5'}$$

After the first braking at station 1 the energy is stored in the energy storage system. During the next acceleration the energy storage system provides the energy to the traction motors. In this case:

$$A_{mech} = A_d * \eta_{gen} + (Ak - \Delta A_{mr}) * \eta_{rech} * \eta_{gen} \tag{6}$$

where: η_{rech} - the efficiency of the energy storage system, operating in the recharge mode.

From (4) and (6) follows:
$$A_{d2} * \eta_{gen} + (A_k - \Delta_{Amr}) * \eta_{ES} * \eta_{rech} * \eta_{gen} = A_k + A_{mr} - \Delta A_{mr} \text{ and :}$$
$$A_{d2} = (A_{mr} - \Delta A_{mr} + A_k)/ \eta_{gen} - A_k * \eta_{ES} * \eta_{rech} + \Delta A_{mr} * \eta_{ES} * \eta_{rech}$$

Before the first braking:
$$A_{mech} = A_{d1} * \eta_{gen} = A_{do};$$

Before second braking:
$$A_{mech} = A_{d2} * \eta gen + (A_k - \Delta A_{mr}) * \eta_{ES} * \eta_{gen} \tag{7}$$

In correspondence with the last two equations:
$$A_{d2} = A_{d1} - (A_k - \Delta A_{mr}) * \eta_{ES} * \eta_{rech} \tag{8}$$

According to equation (8) the use of the energy storage system provides additional energy Δ_{Ad} that equals for the second segment ($n = 2$):

$$\Delta_{Ad} = A_{d2} - A_{d1} = (A_k - \Delta A_{mr}) * \eta_{ES} * \eta_{rech} \tag{9}$$

The above formula shows that the energy gain, achieved by including energy storage system on board is proportional to A_k – the kinetic energy of the vehicle before the braking minus the work of moving resistance forces in the braking mode and the efficiency of the energy storage system.

If we designate $q = A_k - \Delta A_{mr} / A_{mr}$, we receive from (4):

$$A_{mech} / A_{mr} = q + 1 \text{ and:}$$

$$A_k - \Delta A_{mr} = A_{mech} * q / q + 1 \tag{10}$$

It follows from (6) and (10):

$$A_{d2} = A_{d1} (1 - q * \eta_{ES} * \eta_{rech} / q + 1) \tag{11}$$

The formulas (10) and (11) describe a road section with two stations. In case when the section has "n" stations, overall energy consumption for a traditional rail vehicle equals the sum of the energy consumptions in every section between stations.

$$A_{d_{sum}} = n * A_{d_0} \qquad (12)$$

In case of a hybrid rail vehicle, the overall energy consumption taken into account (11) is equal:

$$A_{d_{sum}} = A_{d_0} + (n-1) A_{d_0} (1 - q * \eta_{ES} * \eta_{rech} / q + 1) \qquad (13)$$

The value of the ratio K doc of the overall energy consumption for the hybrid rail vehicle (E_{hrv}) to the energy of traditional rail vehicle (E_{trv}) in the section with "n" station equals:

$$K_{doc} = 1/n + (n - 1/n) * (1 - q * \eta_{ES} * \eta_{rech} / q + 1) \qquad (14)$$

where: K_{doc} – the coefficient of energy consumption.
Let's suppose q = 2, and let's use the formula (14). We can calculate
the value K_{doc} by η_{ES} =0.5, η_{ES} =0.6 and η_{ES} =0.7. The results of the calculation of the value K_{doc} are shown in **Figure 1.9**
If we suppose η_{ES}= 0.6, we can calculate the value K_{doc} by q =9, q =2 and q = 1. The results of the calculation of the values K_{doc} with different q values are shown in **Figure 1.10.**
Let's take a closer look at **Figure 1.9 and Figure 1.10**:
The coefficient of energy consumption decreases in the number of stops (the number of braking episodes).
The efficiency of the energy storage system is crucial to the overall energy consumption efficiency.

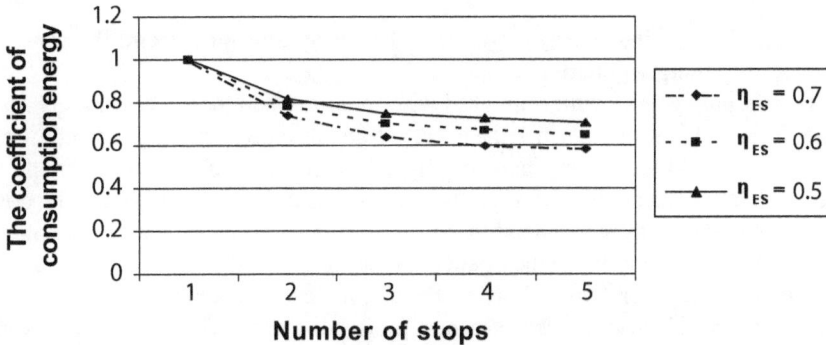

Figure 1.9 Plots of coefficient of decrease of energy consumption as a function of number of the stops and energy storage system's efficiency

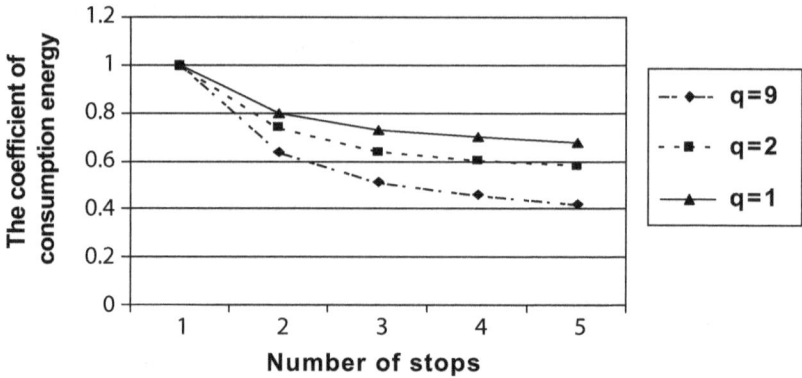

Figure 1.10 Plots of coefficient of decrease of energy consumption as a function of number of stops and the values "q".

By increasing the correlation of kinetic energy to the work of the train resistance forces (the value "q") the gain from implementing energy storage system increases.

The values of the energy efficiency for the converters and electric motors (generators) are very important for energy economy.

Perspective ways to raise the efficiency include:
• implementing of the IGBT transistors as converters
• using motors and generators with permanent magnets
• using battery packs and ultra-capacitors with small internal resistance.

The value q for the section "n" depends on several factors:
• the maximum speed of the traction unit
• the number of grades and curves in the section
• the value of the rolling resistance.

Today's lubricants and friction modifiers decrease rolling resistance between the wheel and the rail, reducing energy consumption for common rolling stock. The application of lubricants and friction modifiers can be also beneficial in the case of hybrid rail vehicles.

1.4.2. The braking energy is stored in the energy storage system; the stops are few, long segments.

Hybrid rail vehicles which are used with trains as main locomotives rarely make stops. Stop braking is rare; hence the energy from this type of braking is insignificant. The electric haul trains mostly benefit from a different braking type: considerable energy economy is accomplished when the speed of the train is stabilized running down a slope. The energy from regenerative braking depends on the mass of the locomotive with the train , the length and the gradient of the slope.

Previously regenerative braking on the diesel-electric locomotives was not used. Today General Electric has started a new production of the diesel-electric locomotives with asynchronous traction motors and energy storage system on board. Usually the diesel-electric locomotives use dynamic braking. In the case of dynamic braking, the electric motors work as generators and the energy from these generators is dissipated as heat in braking resistors. Regenerative braking is similar to dynamic braking. Both types of braking provide advantages such as the decrease in wear and tear on wheels and brake shoes and the possibility to limit the speed in areas with heavy grades. However in the case of regenerative braking energy from braking doesn't dissipate as heat but is stored instead. Particularly in hybrid diesel-electric locomotives this energy is stored in energy storage systems and is used for feeding the auxiliary system and for traction.

Regenerative braking can be used with both dynamic braking and air brakes (electronically controlled or otherwise).

The main formulas that were used in the previous section for stop braking could not be applied to the trains running down the slope. The formulas for this mode are listed below.

If the annual goods that travel in an uphill direction are G1 million gross tons, the annual energy consumption E.C. is equal to:

$$E.C. = 2725*(W_{grade} + W_0 + W_{curve})/\eta_{loc} \text{ [kWh/ (km*year)]}, \quad (15)$$

where: E.C. is the annual energy consumption for the trains moving uphill with the gradient GRAD% per mile by route. Wgrade is the grade resistance (kg/t) W_0 is the rolling resistance (kg/t), Wcurve is the curve resistance (kg/t), η_{loc} is the transmission efficiency.

If the annual goods traffic in the downhill direction is G2 million gross tons, the annual energy return E.R. equals:

$$E.R. = 2725\, G_2*(W_{grade} - W_0 - W_{curve})*\eta_{reg}*\Delta_{reg} \text{ [kWh/(km*year)]}, \quad (16)$$

where: E.R is the annual energy return for the trains lowering in a downhill direction with the gradient GRADE% per mile, η_{reg} is efficiency of the regenerative regime,

$\eta_{reg} = \eta_{ml}*\eta_i*\eta_c*\eta_{ch}$; Δ_{reg} is part of the regenerative braking effort in the total train braking effort by using not only regenerative but also air braking.

The efficiency of the regenerative braking in the slope per 1 km by route in year η_{eff} is equal:

$$\eta_{eff} = E.R/E.C = G_2/G_1*(W_{grade} - W_0 - W_{curve})/(W_{grade} + W_0 + W_{curve})*\eta_{lo}*\eta_{reg}*\Delta_{reg} \quad (17)$$

Therefore, it follows from (14) and (17) that the values of the efficiency of the converters and electric motors (generators) are as important for the economy of the energy as it is by stop braking as by speed stabilization when the train is running down the slope.

If we consider for the average conditions: $W_0 + W_{curve} = 4$ kg/t, [1.2] $\eta_{loc} = 82\%$, $\eta_{ml} = 0.9$; $\eta_i = 0.95$; $\eta_c = 0.95$; $\eta_{ch} = 0.9$, the formula (17) will be simpler:

$$E_{r.b} = 0.6 G_2/G_1*\Delta_{reg}*(W_{grade} - 4/W_{grade} + 4) \quad (18)$$

The diesel locomotives with asynchronous traction motors and energy storage can provide, in the areas with heavy grades, about 7-14% energy economy. This economy will be considerably more if $G_2 > G_1$. As an example, empty trains could travel uphill and after loading they would travel downhill.

1.5 Energy and power of regenerative braking

It is not an easy task to calculate and choose optimal ranges for energy and power of the energy storage systems deployed on different kinds of hybrid rolling stock. It's being complicated by the fact that some of the parameters that should be taken into consideration for example distribution of relative time of work, have a probabilistic nature.

We came up with the results of the calculations of braking energy and power by estimating energy, power and masse of energy storage system. The energy storage system should be capable of storing the energy of regenerative braking. This should become a basic requirement for energy storage systems.

Authors predict that in the near future all rail hybrid vehicles will be adapt to storing and usage of regenerative braking energy.

The results of our calculations for energy and power of regenerative braking for different kind of hybrid rolling stock could be found below. Energy of regeneration braking (EORB) and power of regeneration braking (PORB) are calculated in the vehicle wheel rim. EORB defines the quantity of energy in kW*hours, PORB defines the quantity of energy during the unit of time in kW.

It is necessary to multiply the EORB and PORB by efficiency (correspondence) of components of the energy storage system (ESS). If we multiply EORB and PORB by general efficiency of ESS we will discover the energy and power of an energy storage system.

1.5.1 The hybrid rail vehicles with frequent stops.

The considerable amounts of kinetic energy of the vehicle can be used by a regenerative braking.

In correspondence with formula (5):

$$E_{RB} = (m*v^2/2 - Wo*m*g*S - W_{curve}*m*g*S \tag{19}$$

$A_k = m*v^2/2$, $\Delta A_{mr} = (W_o + W_{curve})*m*g*S$, Wo - the rolling resistance of the locomotive and cars or only railway cars, W_{curve} - the curve resistance of the locomotive and cars or only railway cars, S_b - braking way.

The value PORB can calculate by formula:

$$P_{RB} = E_{RB}/t,$$

where t - time of braking, t = V/b, b - deceleration (m/s²) If ERB is measured in kWh, V - in m/s, and b - in m/c² we can transform this formula to:

$$P_{RB} = 3600\, E_{RB} * b/v\ (kW) \tag{20}$$

The results of the calculation of values ERB and PRB are shown in **Figures 1.11-1.14.**

EORB dependence of speed and weight of vehicle

EORB (kWh)

The speed of rail vehicle (m/s/10)

- -◆- - Weight of vehicle is 50t
- -■- - Weight of vehicle is 100t
- -▲- Weight of vehicle is 250t

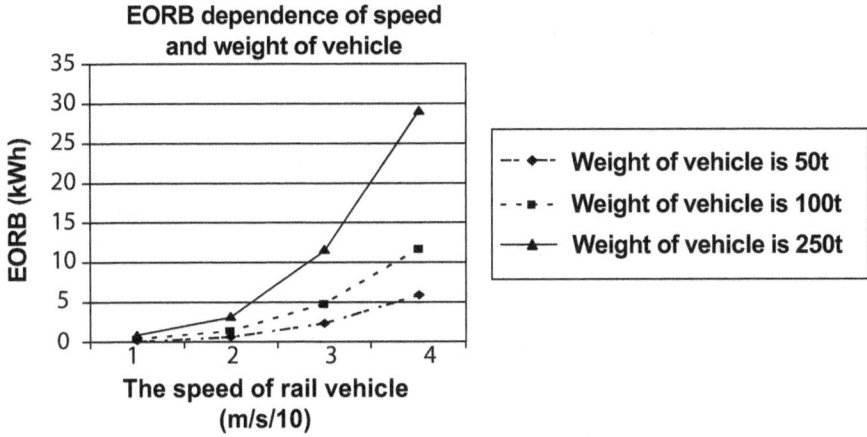

Figure 1.11 EORB for trams, tram-trains, railcars.

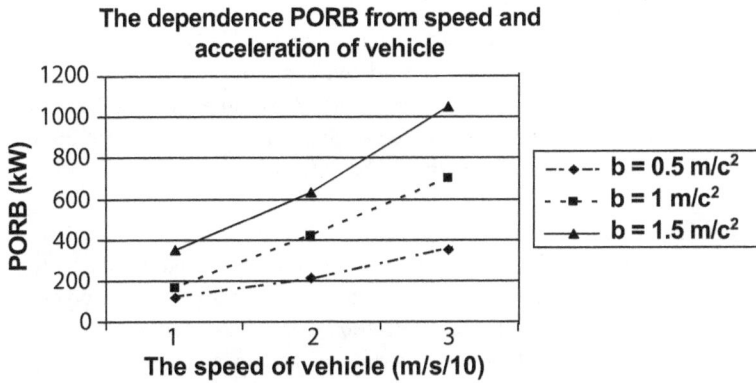

The dependence PORB from speed and acceleration of vehicle

PORB (kW)

The speed of vehicle (m/s/10)

- -◆- $b = 0.5$ m/c^2
- -■- $b = 1$ m/c^2
- -▲- $b = 1.5$ m/c^2

Figure 1.12 PORB for 50t weight vehicle.

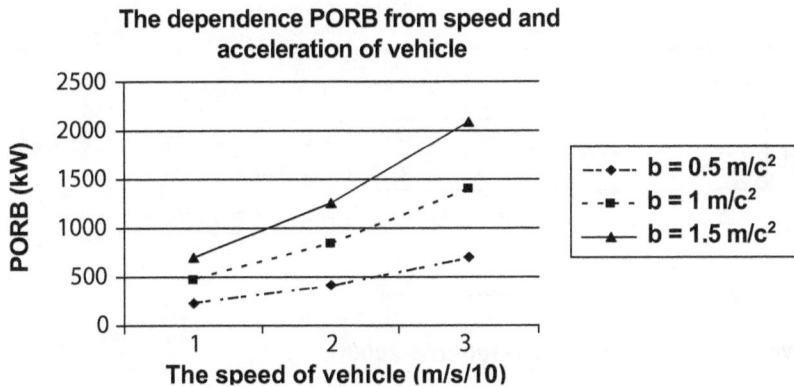

The dependence PORB from speed and acceleration of vehicle

PORB (kW)

The speed of vehicle (m/s/10)

- -◆- $b = 0.5$ m/c^2
- -■- $b = 1$ m/c^2
- -▲- $b = 1.5$ m/c^2

Figure 1.13 PORB for the 100t weight vehicle.

The dependence PORB from speed and acceleration of vehicle

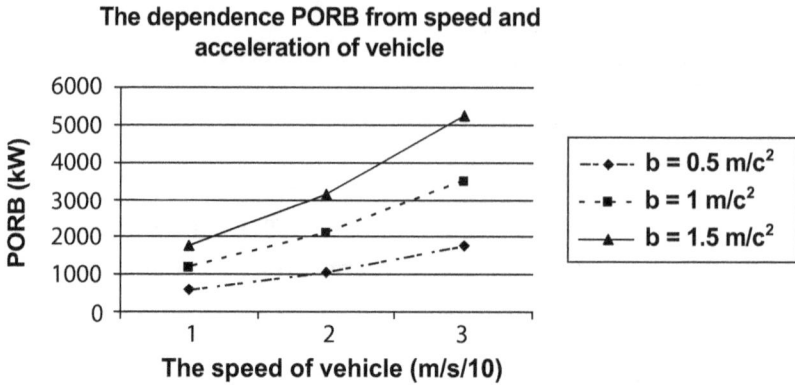

Figure 1.14 PORB for the 250t weight vehicle.

1.5.2 Hybrid switcher locomotive

The characteristics of switcher locomotives depend on the weight of the train, and the frequency of stops. To calculate of the hybrid switcher's EORB we used the formulas (19) and (20). The results of the calculations are shown in **Figures 1.15, 1.16**.

For the switcher locomotive that has a speed of 3m/s and weight of train 2000 t, the EORB equals 1.6 kWh and the PORB should be 960 kW.

1.5.3 Hybrid freight locomotive

Hybrid rail vehicles use the kinetic energy of the train if they make frequent stops which include full stop braking. The use of energy storage systems provides a potential of the regenerative braking to the freight diesel-electric locomotives. Hybrid locomotives use the potential energy of the train if the speed is stabilized on the grades.

Energy of regenerative braking (EORB) of the locomotive and freight train in this case is equal to:

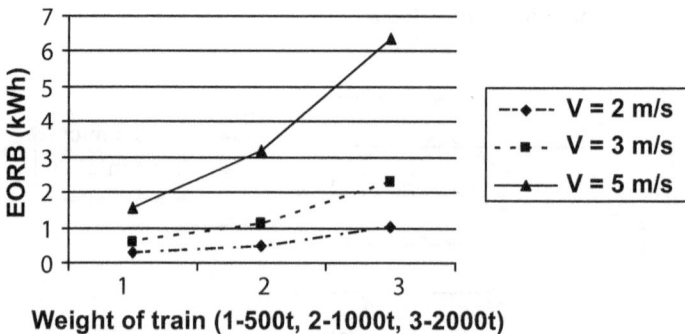

Figure 1.15 EORB of the switcher hybrid locomotives.

PORB dependence of hybrid switcher locomotive with freight train of weight and speed

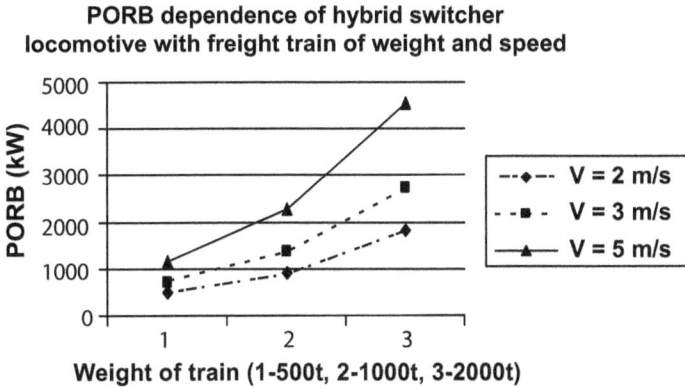

Figure 1.16 PORB for the switcher hybrid locomotives

$E_{RB} = (E_p - W_o * m * g * S_b - W_{curve} * m * g * S_b), E_p$ - potential energy,

$Ep = m * g * W_{grade} * S_b,$ Ep - potential energy, $m = m_{loc} + m_{tr}, m_{loc}$ - masse of locomotive, m_{tr} - masse of the train

$$E_{RB} / S_b = m * g * (W_{grade} - W_o - W_{curve}) \tag{21}$$

The EORB per 1 km by route is proportional to the weight of the locomotive with the train, the value of grade resistance minus rolling and curve resistances of locomotive with the train.

The power of regenerative braking (PORB) is equal to:

$$P_{RB} = E_{RB}/t = (E_{RB} / S_b) * V \tag{22}$$

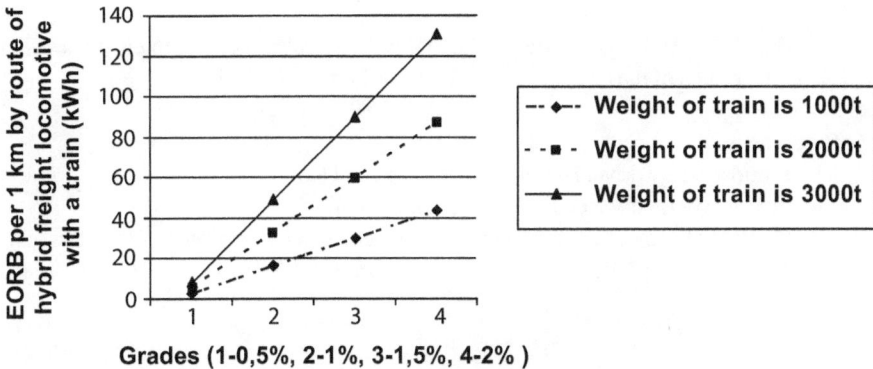

Figure 1.17. EORB per 1 km by route. Hybrid freight locomotive with the train. Grades are different.

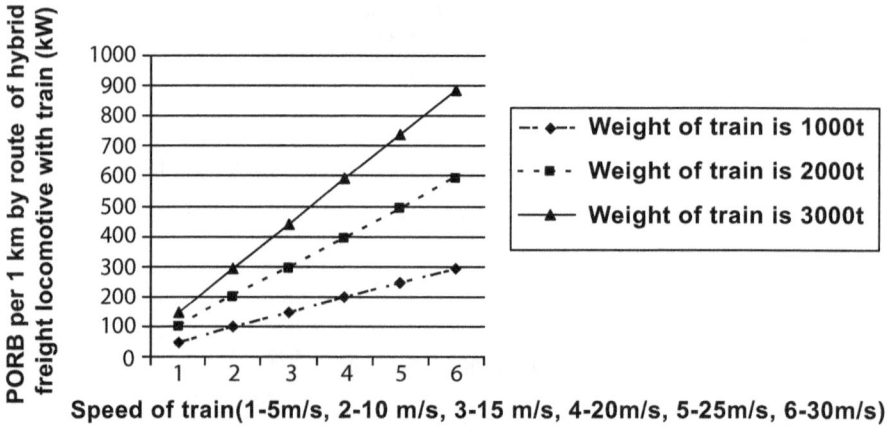

Figure 1.18 PORB per 1 km by route. Hybrid freight locomotive with the train (grade is 0.5%)

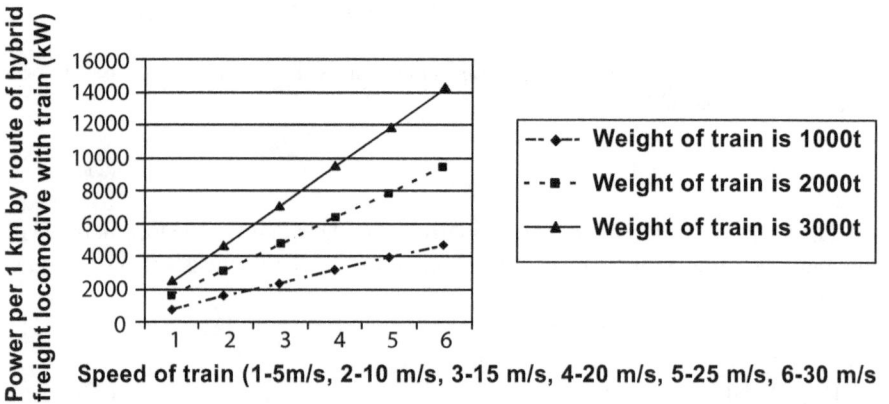

Figure 1.19. PORB per 1 km by route. Hybrid freight locomotive with the train (grade is 2%)

EORB and PORB for hybrid freight locomotive's we calculate by using formulas (21) and (22). The results of the calculation are shown in **Figures 1.17 - 1.19.**

References

1.1. Hybrid synergy drive. Hybrid vehicle: The UN Definition
www.hybridsynergydrive.com/en/un_definition.html - 5k -
1.2. В.Е. Розенфельд, и др. « Теория электрической тяги» Транспорт, 1983 (Russian)
V. E. Rosenfeld et al. "The theory of the electric traction" Transport, 1983 (in Russian)

2. GENERAL ELEMENTS OF HYBRID RAIL VEHICLES

2.1. Prime movers and motors

2.1.1 The diesel engines

In the near future the diesel engine will remain the most important part of the hybrid DEMU and the hybrid locomotive. The diesel engine MTU 16V 4000 R41 [2.1] is an example of a modern diesel engine. This engine is used in the following locomotives: 2016 (Hercules, Austria), ER 20 (Germany), BB475000 (France). The engine is a four-stroke engine with common rail injection. It has two contour systems for liquid cooling, the turbocharger, and an external cooling system for the charger's air.

The common rail injection allows all the fuel injection features that affect fuel combustion to be controlled independently. This includes variables such as the timing period and the method of injection, as well as the injection pressure. These elements are advantageous when it comes to reducing fuel consumption and exhaust emission levels; and this is not just at a specific running speed, but across the entire engine power curve.

The Electronic Diesel Control (EDC) controls all the injection parameters extremely precisely. It incorporates the basic functions for trend analysis, system diagnostics, and has the ability to monitor the engine by remote diagnostics.

In comparison with the diesel engine of the previous generation that powers the 2000 kW diesel 16V 4000 R41, it experiences less fuel consumption and a more favorable relationship of weight to power. The parameters for this new engine are 198 g/kWh and 3.7 kg/kW. In comparison, the parameters for the previous diesel engine were 218 g/kWh and 5.3 kg/kW. The application of the diesel engine 16V 4000 R41 in the diesel locomotive ER 20 allows for a significant decrease in the emissions.

The diesel engine MAN D 2876 LUH 602 of the Lirex DEMU [2.2] also has the turbocharger and external cooling system for the charger's air. The engine of the new generation D2676 has the same UIC-rated power, the common rail injection and fuel consumption 193 g/kWh. Each section of the six-car Lirex DEMU has four diesel–generator sets. All these sets are located on the roof of the cars. The diesel-generator sets of the classic DEMU are commonly located under the floor of the cars. Placing the diesel-generator sets on the roof is a new design concept for the new generation of locomotives.

Sometimes the power unit includes not only a diesel engine, but a generator, and the equipment interconnecting them to make a single pulley (power pack). This design is used for the AGC train.

2.1.2 Microturbines

Microturbines [2.4-2.5] are small gas turbine generators ranging in size from 30 kW to 500 kW. Microturbines can start in less than 30 seconds and offer a number of potential advantages compared to other technologies for small-scale power generation.

These advantages include the small number of moving parts, compact size, low weight, very low emissions, and their long lifetimes. Many different kinds of fuel can be used in a microturbine: natural gas, propane, diesel gasoline and kerosene. Some microturbines use methane gas from landfills and sewage treatment plants.

There are certain advantages of the microturbine over the internal-combustion engine (ICE) and the PEM fuel cell. The lifetime of the microturbine is 20000 hours. This is greater than the lifetime of the ICE which is 12500 hours and the PEM which is 2000 hours. The ICE includes about 50 moving parts, the microturbine instead has only one part, the rotor section. Microturbines don't need pumps, radiators, reservoirs, water, or special anti-freeze.

The turbine system includes a compressor, recuperator, combustor, turbine and permanent magnet generator. The rotating components are mounted on a single shaft supported by air bearings (eliminating the need for lubrication) that rotate at up to 96,000 RPM. The generator is cooled by the air flow into the gas turbine, thus eliminating the need for liquid cooling.

Air is compressed and injected into the recuperator where its temperature is elevated by the exhaust gases expelled from the turbine. This process increases the system efficiency. The heated compressed air is mixed with fuel and burned in the combustion chamber. The combusted hot gases expand through the turbine, providing the rotational power. Patented techniques in the combustion process result in the extremely low emission exhaust stream.

The output of the generator is variable voltage and variable frequency AC power. Power electronics convert this to programmable DC power for hybrid electric vehicle applications.

Capstone Turbine Corporation is the world's leading producer of low-emission microturbine systems. The first power product utilizing microturbine technology was built in 1998. Hybrid buses with Capstone microturbines worked and work now in New Zealand, Tokyo (Japan), Knoxville, Gatlinburg, Chattanooga Indianapolis and Los Angeles (USA). The microturbine system is used for charging the battery pack on the hybrid bus. The buses in Knoxville with NiCd batteries do not have the microturbine system and the range between charges is about 60-90 miles. With the turbine generator as a range extender, the buses have a range of about 250 miles between charges.

2.1.3 Fuel cells

A fuel cell [2.6-2.12] is a high efficiency electrochemical energy conversion device which can generate electricity and produce heat, with the help of catalysts.

The British judge and scientist, Sir William Robert Grove, researched the principal functions of fuel cells in 1839. The next major chapter in the fuel cell story was written by an engineer at Cambridge University, Dr Francis Thomas Bacon. In 1939, he built the alkali electrolyte fuel cell that used nickel gauze electrodes and operated under pressure as high as 3000 psi.

PEM (Proton Exchange Membrane) technology was invented at General Electric in the early 1960s, through the work of Thomas Grubb and Leonard Niedrach. This technology has been especially applicable for transport applications.

In a fuel cell, the chemical energy is provided by a fuel and an oxidant stored outside the cell in which the chemical reactions take place. As long as the cell is supplied with fuel and oxidant, electrical power can be obtained. Therefore fuel cells can work for a long time (about one hundred thousand hours for example). They don't need charging.

The fundamental mechanism of the fuel cell operation is the inverse of the reaction of water electrolysis. Catalytic oxidation of hydrogen at an anode and reduction of oxygen at a cathode create a potential difference between these electrodes. This can be used in an external circuit if an insulating electrolyte between the electrodes allows for ionic mass and charge transfer.

The advantages of the fuel cells with PEM are:
• The solid electrolyte reduces corrosion and management problems
• Low temperature
• Quick start-up

The disadvantages of the fuel cells with PEM are:
• Low temperature requires expensive catalyst
• High sensitivity to fuel impurities

Individual fuel cells are connected in a "stack." The entire operational system consists of a fuel processor or fuel reformer, power conditioning components (such as inverters and voltage controls), motors, compressors, blowers and fans, valves and piping, and even conventional batteries.

The fuel processor is a device for creating relatively pure hydrogen from available fuels by combining fuel refined by hydrogen purification.

The reformer is a fuel cell component that allows hydrogen molecules to be extracted from hydrocarbon fuels via catalysis. Proton exchange membrane fuel cells are energy conversion systems, which directly convert chemical to electrical energy.

The transformation of the energy in the heat engine is different. In this case the chemical energy of the fuel turns to electric energy after several steps. Chemical energy turns first to heat energy. Then heat energy turns to mechanical energy. Only in the third step does the mechanical energy turn to electrical energy. The potential of the efficiency of PEMFC is very large. But practically this efficiency is about 50%.The polarization losses at the electrode interface and the heat losses at the interior resistance decrease the potential possibility. However, there are more losses to be considered. Typically, transport's PEM fuel cells consume 10% or more of the rated stack power output to provide power to pumps, blowers, heaters, controllers etc.

The question about the cost of the PEM stack and system is very important. The cost target of the fuel cell stack and the balance of the plant components for 50 kW systems is $30/kW. This target can be achieved only by manufacturing in a high volume production situation. High volume production is defined as 500,000 units per year. Ways to decrease the cost are:
• Advanced membrane material
• Reduced catalyst loading
• Standardized modular design

According to [2.10] the cost per kW in the year of 1990 was $3000/kW (platinum loading was about 20g/kW), in 2004 this cost was $225/kW (platinum loading was about 0.8 g/kW), and in 2015 year this cost is projected to be $30/kW (platinum loading should be about 0.2g/kW).

The hydrogen is necessary for maintaining the operating condition of the fuel cell. Hydrogen is an energy carrier that can be made from a variety of sources. Today 48%

of the total global hydrogen production is made from natural gas, 30% from oil, 18% from coal, and only 4% from electrolysis. [2.8]

Hydrogen can be made not only from coal, natural gas, biomass, and water electrolysis. It can be made also from common fuels like propane and gasoline, and less common ones like methanol and ethanol, which all have hydrogen in their molecular structure.

Hydrogen is traditionally stored as a compressed gas or as a cryogenic liquid. Both of these storage methods have shortcomings that present problems for the use of hydrogen as a ubiquitous fuel gas.

Transportation industry is actively seeking a new energy storage design. In order to avoid having large, heavy pressure tanks, a liquid fuel is preferable to a gas. Companies are working on a fuel processor for liquid fuels like gasoline and methanol. Methanol is the most promising fuel in the short term; it can be stored and distributed in much the same way as gasoline is now.

Today, rechargeable metal hydrides offer a "solid" alternative to gaseous and liquid storage and are capable of storing large amounts of hydrogen at ambient temperatures and pressures. They can provide a safe, energy efficient, and environmentally sound means of storing hydrogen fuel for use with fuel cells. An additional advantage of metal hydrides is their ability to deliver very pure hydrogen. This is particularly important for PEM fuel cells, which use Pt catalysts that can be easily poisoned if certain impurities (such as CO) are present in the hydrogen. Metal hydrides have negligible, if any, loss during storage, giving them extremely long shelf-life.

The use of hydrogen as a fuel for wide-spread distribution in either gaseous or liquid form poses numerous safety, technical, and economic problems that make its use as a fuel prohibitively difficult. In the absence of a hydrogen pipeline network, small-scale users purchase merchant hydrogen as compressed gas in steel cylinders, or as liquid hydrogen in cryogenic containers. One approach to resolve the drawbacks of hydrogen as a fuel includes considering less expensive, simpler, and cheaper materials that can act as hydrogen carriers.

Ammonia has been identified as a suitable hydrogen carrier. Compared to hydrogen, ammonia therefore offers significant advantages in cost and convenience as a vehicular fuel due to its higher density and its easier storage and distribution. Ammonia is produced and distributed world-wide in million of tons per year. Procedures for safe handling have been developed in every country. Facilities for storage and transport by barges, trucks and pipelines from producer to ultimate consumer are available throughout the world.

Therefore liquid anhydrous ammonia is an excellent storage medium for hydrogen. The fuel capacity per weight of ammonia is higher compared to methanol and the price per kW is lower. Anhydrous ammonia contains 17.8% by weight hydrogen.

Hydrogen can be obtained from coal, natural gas, biomass, and water electrolysis, propane, gasoline, methanol, ethanol, rechargeable metal hydrides, and ammonia. There are many ways to obtain hydrogen. The future will show which option will prove to be the best..

Let's take a brief look at the evolution of the diesel engines. There is increasing demand for newer diesel engines with lower exhaust emmisions and better ecological footprint, though there are potential technology limitations that become apparent as we develop new technologies.

Steam engines have encountered same limitations between the years of 1940-1950. The possibilities of improving the characteristics of steam engines at that time became more and more limiting.

The evolutional stage of development of steam engines has changed with the revolutionary implementation of diesel engines on diesel-electric locomotives in the period of 1950-1960.

In the future we will see if the fuel cell based development of locomotive drive will also become a revolutionary step.

2.1.4 Main generators

The purpose of the main generator [2.13-2.14] in a diesel-electric locomotive is to convert the mechanical horsepower developed in the diesel engine into electrical power.

A synchronous generator consists essentially of a stator and a rotor. The stator core is built up of electrical-sheet steel laminations insulated from one another and is secured within a substantial frame. On its inner surface the stator has slots which receive an AC winding, mostly three-phase. The rotor of a synchronous generator is commonly a salient-pole electromagnet.

Current from an external source, called the exciter, is conveyed to the rotor winding via the slip-rings and brushes. The central shaft of the rotor is coupled to the diesel engine.

The generator is simpler if the rotor has permanent magnets. In this case the generator does not need rotor winding, slip-rings, brushes and exciters. A magnets or electromagnets produce a magnet field. The rotor rotates, and a magnetic field changes. The winding of the stator induces an electromotive force.

If the generator connects to the electric system, the AC current runs in the stator windings. The interaction between the stator currents and the rotor field provides the electromagnetic forces that will tend to pull the rotor against the direction of rotation. This is nothing other than the conversion of the mechanical power available at the output shaft to electric energy. The frequency of the stator's current depends only on the returns per minute of the rotor and the number of stator poles.

The permanent magnet generator is a synchronous machine where the rotor windings have been replaced with permanent magnets. This eliminates the excitation losses in the rotor and increases the efficiency of the generator. The reduced losses also produce a lower temperature rise in the generator, which means that a smaller and simpler cooling system can be used. The temperature reduction in the rotor also reduces the temperature in the bearings, improving reliability by increasing the lifetime of the bearings and the bearing grease. The high efficiency of the generator results in economy of the diesel fuel and decreases the emissions. The reduction in size and weight is also very important for the traction drive.

Recent developments in permanent magnet generator technology have been made possible by a significant improvement in the magnetic materials during the past 15-20 years. A piece of neodymium boron iron (NeFeB) material can have a magnetic force more than 10 times stronger than a traditional ferrite magnet.

The permanent magnet technology is a good design for rail hybrid vehicles. The generator of the DEMU Lirex is an example of such technology. The diesel engine and generator are fastened together at mutual frame places at four points of the car or carriage's roof as one module. On this frame other devices are fastened such as the refrigerator, an air supply system and a system of bent pipe which exhausts the gases.

The generator is a synchronous electric 3-phase machine that exists from the permanent magnets. Its weight and sizes are substantially lesser than by common generators. The rectifier is assembled on the stator's frame that is cooled by the water. It is the common diode bridge. The weight of the generator together with the rectifier is 440 kg which is 40-50% less than that of common generators. The nominal value of the voltage of the generator is 1310 V. The nominal value of the power of the generator is 322 kV*A. The voltage of the generator and the voltage of the rectifier are about proportional to the revolutions of the diesel engine which is regulated in the range of 750-1900 rpm.

2.1.5 Traction motors

Major modern traction rolling stock vehicles use asynchronous traction motors. Hybrid rail vehicles will also use this kind of motor in the near future.

Traction asynchronous motors are 3-phase motors with a squirrel cage rotor. Such motors consist of a stationary part, or the stator, and a rotating part, the rotor. The stator of an asynchronous motor is a hollow cylinder built up by electrical-sheet laminations insulated one from the next.

Usually the stator's windings have a star (Y-shaped) connection. The current flowing in the phase windings sets up a revolving stator field. The number of pole pairs of this field is equal to the number of coil groups in each phase winding.

The rotor core is mounted on a shaft which is supported by bearings. The winding of a rotor consists of solid non-insulated copper or aluminum bar placed in rotors slots. These rotor bars are attached at the two ends of the rotor by end rings to form a structure resembling a squirrel's cage.

The power of the traction motors of the LRV, diesel trains and diesel electric locomotives is generally in the range of 100-650 kW. The ratio power to weight is about 0.25-0.4.

2.2 Converters

2.2.1. The components of the modern traction converters

The components of the traction converters [2.15] are semiconductor devices: diodes, thyristors, and insulated gate bipolar transistors (IGBT).

A diode is a device with two terminals. Its function is to conduct current in one direction and to block its flow in the reverse direction. Diodes allow electricity to flow in only one direction. The current flows if the potential of the anode is greater than the potential of the cathode.

A thyristor is a bistable device comprised of three or more junctions in which at least one of the junctions can switch between reverse- and forward-voltage polarities. The thyristor has three terminals: anode, cathode, and gate. The common thyristor acts as a switch. However, when switched on, it can only pass current in one direction. Each device is turned on at the appropriate time by a trigger pulse applied to the gate and the device will remain on until the instantaneous load current through it drops to zero. The rated "ON" current for some thyristors is as large as 5kA, and the breakdown voltage as high as 5 kV.

The GTO-thyristor (Gate Turn-Off Thyristor) as well as the common thyristor is a switch. GTO is the thyristor, which like the common thyristor with a positive voltage impulse, can be open. In contrast to the common thyristor, the GTO thyristor can be closed with the help of a negative voltage impulse. The GTO was patented in Japan in 1959 but its vast application only began in the late 90s of the 20th century. The rated "ON"current for some GTO is 3 kA and the maximum inverse voltage is 5 kV.

Insulated gate bipolar transistors (IGBT) are bipolar transistors with an insulated gate. They combine the advantages of the bipolar transistor (high voltage and current) with the advantages of the MOSFET (low power consumption and high switching). MOSFET is an acronym for metal-oxide semiconductor field-effect transistor. The IGBT is a recent invention. The "first-generation" devices of the 1980s and early '90s were relatively slow in switching. The second generation devices were much improved, and the current third generation ones are even better, with speed rivaling MOSFETs, excellent ruggedness and tolerance to overloads. The extremely high pulse ratings of the second and third generation devices also make them useful for generating large power pulses in area converters for traction drives. In 2000 the breakdown voltage rate of the IGBT was 6.5 kV and the current was 600 A. Those very high parameters of the modern IGBT allow it to be used in nearly every converter of hybrid rail vehicles.

2.2.2 3-phase Rectifiers Bridge

In the diesel-electric vehicles, the 3-phase rectifier bridge is commonly used between the main generator and the DC-link of the PWM converters. This rectifier has two groups of the diodes, an anode and a cathode. Every group has three diodes. The diodes of the anode and cathode group form a 3-phase bridge. The positive pole is the general point of the cathode group; the negative pole is the general point of the anode group.

Three sinusoidal voltages are at the input of the rectifier. The phases of voltages are spaced apart by an angle at a 120 degree electric grade. At the output of a rectifier is the DC voltage with a small 6-phase ripple. The filter in at the input of the DC link deletes the ripple. A diagram of diesel-generator power source with a schema of the rectifier is shown in **Figure 2.1.**

2.2.3. The choppers

Several hybrid rail vehicles today use DC drive. They have DC motors and choppers [2.16] Furthermore the choppers are used as buck-bust converters in hybrid rail vehicles between the DC link and the energy storage units with the battery bank or the ultracapacitors.

The buck and boost chopper with the IGBT transistor is shown in **Figure 2.2.**

This is a converter for a hybrid rail vehicle. The energy storage uses ultracapacitors. The traction motors are fed from the DC bus through the PWM inverter. When

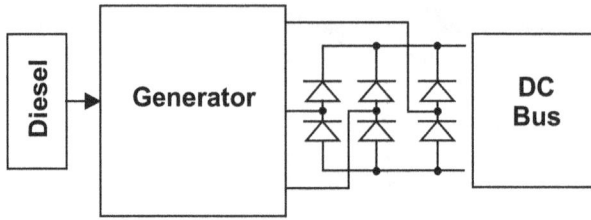

Figure 2.1 Diagram of diesel-generator power source with schema of the rectifier

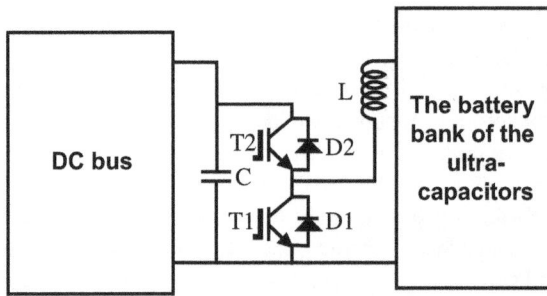

Figure 2.2. Block diagram of energy storage with ultracapacitors

the T1 transistor is closed, the voltage from the energy storage through the diode D2 goes to the DC bus and to the power inverter (mode of acceleration). When the transistor T1 is opened, the current of the energy storage continues to flow through this transistor. The shopper works as a boost (step-up) converter. When the T2 transistor is open the chopper works as a buck (step-down) converter. The voltage from the traction motors these work as the generators go to DC bus and then to the energy storage (mode of regenerative braking). When the T2 transistor is closed, the current from the energy storage continues to flow through the D1 diode.

In hybrid rail rolling stock the chopper works as a step-up converter in traction mode (acceleration mode) and as step-down in braking mode.

2.2.4. PWM inverter.

The inverter is fed by a DC voltage and has three phase legs each consisting of two transistors and two diodes **(Figure 2.3)**. There is a sequence of 3 impulses in the exit of the inverter. These 3 sequence impulses are spaced apart by a 120 degree electric grade.

The curves of an output voltage unit using the control method are sine-triangle pulse width modulation controls. With such a control, the switchers of the inverter are controlled based on a comparison of a sinusoidal control signal and a triangular switching signal. The sinusoidal control waveform establishes the desired fundamen-

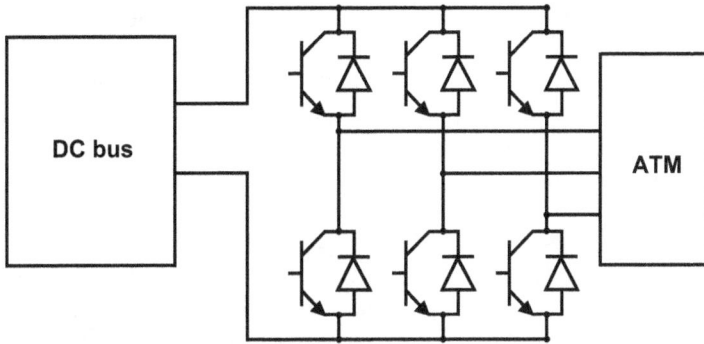

Figure 2.3 Block diagram of asynchronous traction drive with PWM inverter

tal frequency of the inverter output, while the triangular waveform establishes the switching frequency of the inverter.

The 3 different transistors in the main power circuit are "On" at all times. The output terminals of the voltage-source inverter for phases a, b, c periodically connect to positive and negative poles of the feeding source and are characterized by an impulse sequence.

In most instances the magnitude of the triangle wave is held fixed. The amplitude of the inverter output voltages is therefore controlled by adjusting the amplitude of the sinusoidal control voltages. The ratio of the amplitude of the sinusoidal waveforms relative to the amplitude of the triangle wave is the amplitude modulation ratio. In systems in which the inverter sources inductive loads, the inverter must be the source of power in all four quadrants. The asynchronous motor is an example of inductive loads. The diodes provide paths for current when a transistor is gated on but cannot conduct the polarity of the load current. For example, if the load current is negative at the instant the upper transistor is gated on, the diode in parallel with the upper transistors will conduct until the load current becomes positive at which time the upper transistor will begin to conduct.

2.3. Modern energy storage devices
2.3.1. Battery Packs
2.3.1.1 Lead Acid Battery

The lead-acid battery [2.17-2.18] was invented in 1859 by French physicist Gaston Planté.

Lead acid batteries use lead oxide (PbO_2) positive electrodes, lead (Pb) negative electrodes, and about 37% sulfuric acid (H_2SO_4) as the electrolyte.

The overall chemical reaction for the lead acid battery is:

$$Pb + PbO_2 + 2H_2SO_4 \longleftrightarrow 2\,PbSO_4 + 2\,H_2O$$

During of the charge and recharge the concentration of the acid changes. The main characteristics of the lead acid battery are listed in **Table 2.1.**

Modern lead acid batteries are maintenance free as well as the VRLA and AGM batteries.

The construction of the Valve Regulated Lead Acid (VRLA) batteries is designed to prevent electrolyte loss through evaporation or spillage and this in turn prolongs the life of the battery and eases maintenance. Instead of simple vent caps on the cells to let gas escape, VRLA have pressure valves that open only under extreme conditions. Absorbed glass mat (AGM) is a class of lead-acid batteries in which electrolyte is absorbed into a fiberglass mat. The plates in an AGM battery may be flat like a wet cell lead-acid battery, or they may be wound in a tight spiral. The fiberglass mat absorbs and immobilizes the acid in the mat but keeps it in a liquid rather than a gel form. In this way the acid is more readily available to the plates allowing faster reactions between the acid and the plate material allowing higher charge/discharge rates as well as deep cycling.

This construction is very robust and able to withstand severe shock and vibrations and the cells will not leak even if the case is cracked.

Table 2.1 Main characteristics of the modern batteries

Battery Type	Energy Density Wh/Kg	Specific Power W/Kg	Cycle Life Cycles	Average Energy Cost $/kWh
Lead Acid	30-40	200-300	500-800	150
NiMH	30-80	250-1000	1500	250
ZEBRA	100-120	150	1200	300
Lithium-Ion	150	800-1500	1000-1200	300
Lithium-ion Polymer	150	720	1100	160

AGM batteries are also sometimes called "dry," because the fiberglass mat is only 95% saturated with sulfuric acid and there is no excess liquid.

Nearly all AGM batteries are sealed valve regulated "VRLA."

AGM's have a very low self-discharge rate of from 1% to 3% per month.

Advantages of Lead Acid batteries.
• Low cost.
• Reliable. Over 140 years of development.
• Robust. Tolerant to abuse and overcharging.
• Low internal impedance.
• Many suppliers world wide.
Shortcomings of Lead acid batteries
• Very heavy and bulky.
• Loss of power capacity at low temperature.
• Sensitive to mishandling.
• Typical cycle life 300 to 500 cycles.

2.3.1.2. Nickel Metal Hydride (NiMH) Battery
Stanford Ovshinsky, founder of Ovonic Co, earned a patent in 1994 for a high energy-storage, environment-friendly, maintenance-free, rechargeable battery. This was

the Nickel Metal Hydride Battery [2.17-2.18]. This battery is in many ways a development of the nickel–cadmium battery, although they're also related to the hydrogen–nickel oxide batteries. They basically evolved from the work done in the 1970s in the storage of hydrogen gas in metallic hydrides. Like nickel-cadmium, iron-nickel, and nickel-zinc, NiMH batteries use a nickel/nickel hydroxide positive electrode and potassium hydroxide as the electrolyte.

However instead of a cadmium/cadmium hydroxide, iron/ iron hydroxide, and zinc/zinc hydroxide negative electrodes, the NiMH has an electrode made from a hydrogen storage alloy such as lanthanum-nickel. Lanthanum is an element of the rare earth group of the metals of atomic number 57, allied to aluminum. Lanthanum was named after the Greek word lanthanein (to lie hidden).

The overall chemical reaction for the NiMH battery is:

$$NiOOH + MH \longleftrightarrow M + Ni(OH)_2$$

The main characteristics of the NiMH battery are listed in Table 2.1.
Advantages of NiMH batteries:
• The energy density of the NiMH battery is more than double that of the lead acid battery and 40% higher than that of NiCad's.
• The specific power of the battery is high.
• Using NiMH batteries, up to 3000 cycles at 60% Depth of Discharge (DOD) and 2000 cycles at 80% DOD have been demonstrated. At lower depths of discharge, for example 4% DOD, more than 350.000 cycles can be expected.
• Low internal impedance and flat discharge characteristics.
• The ability of fast recharging (the recharge time is 1 h, rapid charge to 60% capacity is 20 minutes).
• Good performance at low temperatures.
Shortcomings of NiMH batteries.
• Very high self-discharge rate.
• Less tolerant of overcharging than NiCad's.
• Cell voltage is only 1.2 Volts which means that many cells are required to make up high voltage batteries.

2.3.1.3. ZEBRA (Sodium/Nickel – Chloride) Battery
In the year 1980, patents were issued for the first ZEBRA Sodium/Nickel Chloride cell. They originated in the mid 70's by the Council for Scientific and Industrial Research (CSIR) in South Africa. It was finally developed and patented by the UK Atomic Energy Authority in Harwell. ZEBRA is an abbreviation for Zero Emission Battery Research Activity.

ZEBRA batteries [2.17-2.18] use the sodium and nickel-chloride electrodes and a beta-alumina (boehmite) ceramic electrolyte.
The overall chemical reaction for the ZEBRA battery is:

$$2Na + NiCl_2 \longleftrightarrow 2NaCl + Ni$$

The battery has to work at high temperatures (270 degree C).

However, this problem is already being managed by the special Battery Management System (BMS) that allows for the use of the battery under very different ambient temperatures, from -40 degree C to +70 degree C. The chemistry of Sodium Nickel Chloride allows for a nominal operation of the cell voltage at 2.58 Volts.

As of the latest reports, the ZEBRA battery is said to have achieved more than 60,000 miles in one vehicle with no maintenance and more than 5 years of life in a car.

The main characteristics of the ZEBRA battery are listed in **Table 2.1.**

Advantages of ZEBRA batteries:
• High specific energy (4 times higher than Lead Acid).
• Low cost materials (ZEBRA requires only 1.53 kg of nickel to produce 1 kWh of energy while Nickel Metal Hydride requires between 3.5 and 6.8 of nickel per kWh).
• Proven safety.
• Performance independent of ambient demonstrated in arctic and desert condition.
• Tolerant of short circuits.

Shortcomings of ZEBRA batteries:
• High internal resistance.
• High operating temperature.
• Preheating needed to get battery up to 270 degree C operating temperature.
• Uses 14% of its own capacity per day to maintain temperature when not in use.
• Thermal management needed.

2.3.1.4. Lithium Ion Battery

Lithium Ion rechargeable batteries are a recent development from the lithium primary cell, which was invented in 1912 but not marketed commercially until the early 1970s. The first practical rechargeable Li-ion battery was developed by Sony in the late 1980s and marketed in 1990. [2.17-2.19]

Lithium batteries consist of positive electrodes, made with lithium metal oxides, negative electrodes, made with carbon materials, electrolytes, composed of organic solvents and lithium salts, and separators. The examples of lithium metal oxides are lithium cobalt $LiCoO_2$, lithium manganese $LiMn2O_4$, and Lithium Nickel $LiNiO_2$.

The overall chemical reaction for the Lithium Ion battery is:

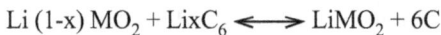

$$Li (1-x) MO_2 + LixC_6 \longleftrightarrow LiMO_2 + 6C$$

The main characteristics of the Lithium Ion battery are listed in Table 2.1.

Advantages of Lithium Ion batteries:
• High specific energy.
• High specific power.
• Fast charge possible.
• High cell voltage of 3.6 Volts means fewer cells and associated connections and electronics are needed for high voltage batteries.

Shortcomings of Lithium Ion batteries:

• High internal impedance.
• High cost of batteries for high power applications.

2.3.1.5. Lithium Ion Polymer Battery

The lithium ion polymer battery, or more commonly lithium polymer batteries are rechargeable batteries, which have technologically evolved from lithium ion batteries. Lithium ion polymer batteries started appearing in consumer electronics around 1996. The electrodes of the lithium ion and the lithium polymer batteries are same. Electrolyte of lithium polymer battery is solid conductive polymer with plasticizer and lithium salt.

The overall chemical reaction of the Lithium Ion polymer battery is:

$$xLi + M_yO_z \longleftrightarrow Li_xM_yO_z$$

If metal is cobalt, reaction is:

$$Li + CoO_2 \longleftrightarrow LiCoO_2$$

If metal is nickel, reaction is:

$$Li + NiO_2 \longleftrightarrow LiNiO_2$$

If metal is manganese, reaction is:

$$Li + Mn_2O_4 \longleftrightarrow LiMn_2O_4$$

The main characteristics of the Lithium Ion Polymer battery are listed in **Table 1.**

The voltage of a lithium polymer battery cell varies from about 2.7 V (discharged) to about 4.23V9 (fully charged).

Advantages of Lithium Ion Polymer batteries:
• High specific energy.
• No material constraints.
• Low service demand.
• Potentially low cost.
Shortcomings of Lithium Ion Polymer batteries:
• Cycle life needs to be improved.

2.3.2. Ultracapacitors.

Like batteries, ultracapacitors are energy storage devices. The ultracapacitors [2.20-2.22] also are known as the supercaps and the double-layer capacitors. The ultracapacitor has the electrodes and the electrolyte as a battery. It is an electrochemical device. But the batteries store charges chemically, whereas ultracapacitors store them electrostatically like the common capacitors.

Ultracapacitors are simply capacitors employing plates (electrodes) with extremely high surface areas providing a high storage capacity. Activated carbon is ideally suited as an electrode. The electrolyte, the conductive liquid between the electrodes, is a conducting salt dissolved in an aqueous or organic solvent.

The capacitance of the ultracapacitor is in proportion to the area of the parallel plates and is inversely proportional to the distance between them.

Maximizing the surface area of the electrodes within the available space means the thickness of the dielectric must be minimized. The thickness of the dielectric is accomplished with nanodimentions. The electrolyte can have a voltage of 2V or in special cases, 3V.

The ultracapacitor with a voltage of 2.5V and a capacitance of 3500 F today has an energy density of 5-10 Wh/Kg, specific power of 6-8 kW/kg, a cycle life of cycles > 100000, and an average energy cost of 580 $/kWh.

A paper presented by Bombardier in June 2006 at the World Congress on Railway Research in Montréal showed that the ultracapacitor's energy storage density had more than tripled since 1998 to reach 6 Wh/kg. It is very important to note that the maximum power over the same period has increased seven fold to 6 kW/kg.

Advantages of Ultracapacitors:
• Very high specific power.
• Fast charge and discharge capabilities.
• Very high energy efficiency.
• Excellent temperature performance.
• Excellent cycle life.
Shortcoming of Ultracapacitors
• Low specific energy.

Using an ultracapacitor in conjunction with a battery combines the performance of the former with the greater energy storage capability of the latter. It can extend the life of a battery, save on replacement and maintenance costs, and enable a battery to be downsized. At the same time, it can increase available energy by providing high peak power whenever necessary.

2.3.3. Flywheels

Flywheels [2.23-2.24] are kinetic energy storage devices, and store energy in a rotating mass (rotor), with the amount of stored energy dependent on the mass and form, and rotational speed of the rotor. A new application of flywheels is in the storage of electrical energy, which is achieved by the addition of an electrical machine and power converter. The electrical machine may be integrated with the flywheel, and operate at variable speed, and the power converter is usually provided by a power electronic variable speed drive.

Traditional flywheel system had a steel flywheel. Steel allows designers to keep manufacturing costs down while maintaining adequate factors of safety. The rotational speed was 2000-3000 rpm. The flywheel systems with the same low rotational speeds enable the use of conventional bearing systems. Traditional flywheels usually operate in air, which causes increased aerodynamic drag losses as well as a higher operating noise level. The integration of an external flywheel requires multiple bearing sets, which can reduce overall system reliability and increase maintenance costs.

Advantages of traditional flywheels are: steel as material is safe and predictable, low RPM makes design simple, inexpensive materials keep costs down. Disadvantages of traditional flywheels are: low energy and power density, multiple bearing sets, high aerodynamic noise and drag.

High speed flywheels (25000-80000 RPM) are made of composite material.

High speed storage systems have relatively low weight and dimensions.

To reduce friction, the rotor runs in a vacuum and is supported by a magnetic bearing. A simplified view of a Flywheel Energy Storage System (FESS) is shown in **Figure 2.4.**

A motor-generator is usually a high speed permanent magnet machine, integrated with the rotor. This is usually known as an integrated synchronous generator. The power electronics interface is usually a pulse width modulated (PWM) bi-directional converter using insulated-gate bipolar transistor (IGBT) are: low energy density, high material costs and expensive magnetic bearing .technology. This converter is similar in its schematics as the inverter in **Figure 2.3 .**

Steel rotors have specific energy up to around 5Wh/kg/ while high speed composite rotors have achieved specific energy up to 100Wh/kg. Specific power figures up to around 1600 W/kg have been quoted. However the specific energy and power of the complete system may be reduced by at least a factor of 10 when the weight of the complete system, including containment, vacuum system, and electrical interface, is taken into account.

The high cycling capability of flywheels is one of their key features. Full-cycle lifetimes quoted for flywheels range from in excess of 10^2, up to 10^7. The highest cycling lifetimes would only be exceeded after 20 years with continuous cycling at the rate of one full charge-discharge cycle every 100 minutes.

Advantages of the high-speed flywheels are:
• compact,
• high efficiency,
• low maintenance,
• no aerodynamic noise.

Shortcomings of the high-speed flywheels are:
• low energy density,
• high material costs,
• expensive magnetic bearing.

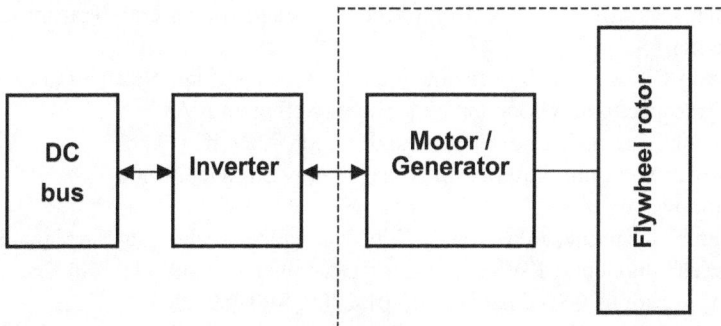

Figure 2.4 Block diagram of a flywheel energy storage system

References

2.1 Marl et al. "The diesel locomotive Hercules with electric drive" Glasers Annalen, 2001, # 6/7, S. 213-222.

2.2 A.Kattner "The diesel engines of MAN AG in Railways" Glaser Annalen, 2000, # 2/3, S. 193-199

2.3 Roy J. Primus "GE Evolution Locomotive Diesel Engine" 11 th Diesel Engine Emissions Reduction Conference. Chicago, Illinois August 24, 2005.

2.4 Green Car Congress. 28 September 2004. Propane Series-Hybrid Buses for Knoxville and Gatlinburg.

2.5 Capstone Turbine-Application.www.microturbine.com/applications/hevApp-Details.asp

2.6 Toyota fuel cell hybrid vehicle (FCHV) book. Toyota Motor Corporation. March 2003

2.7 Franz Albert Horl, Franz Klier, Dirk Leinhos "Brennstoffzellen fur Bahnanwendungen" Eisenbahn Technische Rundschau (ETR) 2001, # 9. S.532-540.

2.8 Karim Nice "How Fuel Processor work"
http://auto.howstuffworks.com/fuel-processor.htm

2.9 Hydrogen Fuel Initiative. Department of energy USA.James F.Miller Director, Electrochemical Technology program Argonne National Laboratory. 8 November 2005.

2.10 Aymeric Rousseau, Phil Sharer, Ragesh Ahluwalia "Optimization of fuel cell vehicle fuel economy" Argonne National Laboratory
www.transportation.anl.gov/pdf/mc/331.pdf

2.11 Green Car Congress. 22 March 2005. New Materials for Hydrogen Storage.

2.12 Ali T-Raissi "Hydrogen from Ammonia and Ammonia-Borane Complex for Fuel Cell Application" (Proceedings of the 2002 U.S. DOE hydrogen Program Review NREL/CP-610-32405
www1.eere.energy.gov/hydrogenandfuelcells/pdfs/32405b15.pdf

2.13 A.S.Kasatkin and M.V.Nemtsov "Electrical Engineering". English translation, Mir Publishers, 1990.

2.14 D.Lenhard et al. "Elektrische Ausrustung des triebzuges LIREX- Baureihe 618/619 fur DB Regio"Elektrische Bahnen. # 8, 2000, s.279-287.

2.15 Бурков А.Т. «Электронная техника и преобразователи» Транспорт 2001 (Russian) Burkov A.T. "Electronic techniques and converters". Transport, 2001 (in Russian)

2.16 Juan Dixon, Micah Ortuzar, Eduardo Arcos and Ian Nakashima" ZEBRA plus ultracapacitors: A good match for energy efficient EV's"
http://www2.ing.puc.cl/power/paperspdf/Dixon/72a.pdf

2.17 Power Solution. Batteries and other Energy Storage Devices.
www.mpower.com

2.18 James Larminie, John Lowry "Electric Vehicle Techno;ogy Explained" Part "Batteries" Publicher: J.Wiley & Sons (December 19, 2003) Hybrid Electric Vehicle Applications", SAE paper 2003-01-2289, SP 1789,2003.

2.19 S. M. Lukic, S. Al-Hallaj, J. R. Selman, and A. Emadi, "On the suitability of a new high-power lithium ion battery for hybrid electric vehicle applications, "Society of Automotive Engineers (SAE) Journal, Paper No. 2003-01-2289, SP-1789, 2003;

2.20 National Renewable Energy laboratory "Advanced Vehicle and Fuels research; Energy Storage, Ultracapacitors"
http://www.nrel.gov/vehiclesandfuels/energystorage/ultracapacitors.html

2.21 P.Barrade,A.Rufer "Current capability and power density of supercapacitors; considerations on energy efficiency, EPE 2003; European Conference on Power Electronics and Application, 2-4 September, Toulouse, France.

2.22 Green Car Congress. 7June 2005. Maxwell Introduce Higher Voltage, Longer-Lasting Ultracapacitors.

2.23 Gerhard Reiner. Schwunggrad-Energyspeicher. Munchen, 14. Marz 2006
http://www.vde.com-NR-rdonlyres-EFCD5ABF-42F2_4093-AF7C-865...

2.24 E.R.Furlong, M.Piemontesi, Prasad P. and Sukumar De "Advances in energy storage techniques for critical power systems" The Battcon 2002 proceeding P. 1-8

3. CONVENTIONAL AND DUAL-MODE ROLLING STOCK

3.1 Dual-mode tram-trains

We have seen many dual-system and dual-mode system trams developed in the recent years. In general, they are known as tram-trains. [3.1-3.4]. A tram-train is a light rail transport system within which trams are able to run on both tram and train tracks for greater flexibility and convenience. With this new system, passengers traveling from outside of a city don't have toz change from train to tram at a central station.

A tram-train is a vehicle that is designed to operate on a city light rail or tram network, normally powered at (600-750)V DC, as well as on a Main Line Railway powered at a different voltage level of 1500V DC, 15 kV, 16.7 Hz or 25 kV, 50 Hz . The vehicle has to comply with two different sets of standards for urban tram/light rail networks and Main Line Railways.

The first section of the dual-system tram-train road opened in Karlsruhe (Germany) in September 1992 and by now many cities across Europe have adopted the same approach. During the period from 1993 till 1997 thirty-three cities in Europe investigated the possibility of using the first generation of tram-trains. Later another forty European cities have adopted the second generation of tram-trains.

Today we have three major families of tram-trains: Regio Citadis (Alstom), Combino (Siemens) and Avanto (S70) (Siemens). Every family has a variable traction system and includes:

- a pure tram system
- a dual-system electric/electric tram-train that can operate on both tram network and the regional electric rail network (electric/electric: powered at 600/750V DC or 15 kV 16.66 Hz; 25 kV, 50 Hz; 1500V DC;)
- a dual-mode electric/diesel tram-train system that can operate on both tram network and the regional not powered rail network (electric/diesel)

The main characteristics of European tram-trains from those families are listed in **Table 3.1.**

The main power circuits of the dual-mode Regio Citadis system are shown on **Figure 3.1.** This tram-train can work in both diesel and electric mode. In electric mode, the switch is in ON position and traction motors are fed through the inverters from DC overhead line with the voltage of 600 or 750V. In diesel mode, the switch is OFF and every two traction motors are fed from separate diesel engines through main generators with permanent magnets, the rectifiers and inverters. The voltage of the DC link in the diesel mode is 450-930 V DC.

3.2 Diesel and dual-mode trains

The majority of the modern diesel trains are made as the Diesel Multiple Unit (DMU). [3.5-3.9]

The DMU classification is subdivided by the form of transmission used:
- DMMU (Diesel Mechanical Multiple Unit) has a mechanical transmission;
- DHMU (Diesel Hydraulic Multiple Unit) has a hydraulic transmission;
- DEMU (Diesel Electrical Multiple Unit) has an electrical transmission.

The examples of the modern diesel trains are listed in **Table 3.2.**

Table 3.1 European tram-trains - main characteristics

Name of tram train	Regio Citadis	Avanto, Paris	Combino, duo
Wheel Arrangement	Bo'- 2'- 2'- Bo'	Bo'- 2'- 2'- Bo'	Bo' - Bo'
Vehicle empty weight, t	62	59.7t	25t
The sum of traction motors powers, kW	600	800	400
Seats/Folding seats	90/10	80/6	27
Maximum speed, km/h	80	100	70
Track gauge, mm	1435	1435	1000
Minimum Curve radius, v	23	20	15
Wheel diameter new, mm		660/610	600
Vehicle length over couplers, mm	36,752	36,965	20.048
Vehicle width, mm	2650	2650	2300
Maximal vehicle height (above TOR), mm	3650	3520	3510
Traction motor converter	2 IGBT PWM Inverter	2 IGBT PWM Inverter	2 IGBT PWM Inverter
Diesel engine power, kW	375, MAN, 6 cylinder	No	180,Euro 3
The ratio of the vehicle power to the vehicle's empty weight, kW/t	9.68	13.4	16
The ratio of the vehicle empty weight to the number of seats, kg	620	694	926

The number of DEMU trains is limited, but we have great interest in DEMU trains as their design resembles the design of the Hybrid Diesel Multiple Unit (HDMU). HDMU train, on the other hand, is DEMU train with additional energy storage.

DEMU is a combination of DMU (Diesel Multiple Unit) and EMU (Electrical Multiple Unit) technologies. DEMU trains are similar to EMU trains, but use power, supplied by an onboard diesel engine.

In most modern DEMU trains, each car is entirely self-contained and has own engine, generator and electric motors. Many modern DEMU trains have dynamic braking, which is accomplished by switching the motors to act as generators that convert motion into electricity instead of electricity into motion. The energy produced

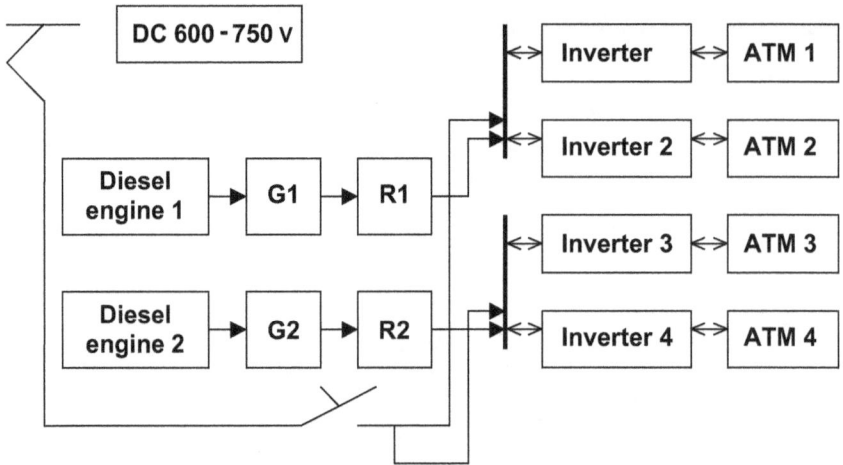

Figure 3.1 Main power circuits - dual-mode variant - Regio Citadis dual-mode tram.

is usually emitted as heat via resistors to the atmosphere. The principle reason for this configuration is to reduce wear and tear on the main service brakes.

Most modern DEMU trains are built on the standard platform of train families. Some of these families are known as Desiro, Talent, GTW, Coradia Lirex. These train families have modern, comfortable and economical cars ideally suited to provide quick connections between regions and cities. Thanks to its modularity it offers many possibilities for individual adaptation of the vehicle and the interior.

The majority of the modern DEMUs has articulated architecture, asynchronous traction drive with IGBT converters, attractive designs both inside and outside, and can be designed according to the client's desire. They are light weight cars.

A good example of the modern family of DMU/EMU/DEMU trains is the AGC (Autorail Grande Capacité) regional train (France). This family of trains was built by Bombardier.

The AGC represents the new generation of regional trains designed to meet current needs for the development of urban and intercity rail transportation. The AGC is available in numerous versions or models. The seating capacity of the trains can range from 160 to 220 seats, depending on the number of cars. Modular interior designs exist in the High Class and Intercity versions. It is a dual-mode system. The trains can run on either diesel fuel, electricity or a combination of the two.

The AGC will travel at 160 km per hour. Thanks to its articulated architecture, it sports wide car bodies and inter-circulation gangways, as well as a continuous low floor. These features provide excellent access for travelers, make it easier to move about in the trains and deliver greater comfort, visibility and security. The AGC not only meets the expectations of mass transit operators and users, but also respects environmental requirements and operating economics.

A block diagram of power circuits on a train in the diesel mode is shown in **Figure 3.2**.

• Traction converters use IGBT.
• General element for all variants of the traction drives is the 1500 B DC bus.
• Modules of the traction converters and a converter of the auxiliary equipment are fed from this bus.
• Rectifier R is fed from asynchronous generator G.
• Diesel engine turns the generator.
• Traction motors M are fed from traction converters.
• Converter includes the braking chopper for the regulation of the braking resistor's current.

The AGS series consists of 4 models:
• X 76500: diesel variant, also designated XGC
• Z 27500: electric variant capable of running on both 1.5 kV DC and 25 kV 50 Hz AC, also designated ZGC
• B 81500: dual-mode variant, capable of running on both diesel (by means of a diesel-electric engine) and 1.5 kV DC (by means of a pantograph), also designated BGC
• B 82500: dual-mode/dual-system variant, capable of running on both diesel (by means of a diesel-electric motor) and 1.5 kV DC/25 kV AC (by means of a

Table 3.2 Main characteristics of modern diesel trains

	Series/Country	First Year in service	Maximum speed (km/h)	Mass tones	Diesel rating, kW	Seats
DEMU	Bombardier Talbot 643 (Germany)	2000	120	86.5	2*500	161
	Stadler/Bombardier DWA 646 (Germany)	2000	120	54.5	1*550	118
	Lirex 618 (Germany)	2002	160	137	2*315	129
	Tilting train VT610 (Germany)	1992	160	95.35	2*485	136
	XGC Bombardier (France)	2004	160	136.5	2*662	160
DMMU	RegioSprinter (Germany)	1985	120	31.9	2*228	80
	Agenda BM93 (Norway)	2000	120	90	2*306	92
DHMU	Colorado Railcar DMU commuter railcar (USA)	2005	160	144	2*441	188
	X 72500 TER Alstom (France)	1997	160	116	2*600	150
	Desiro 185 (United Kingdom)	2005	160	168	3*561	168

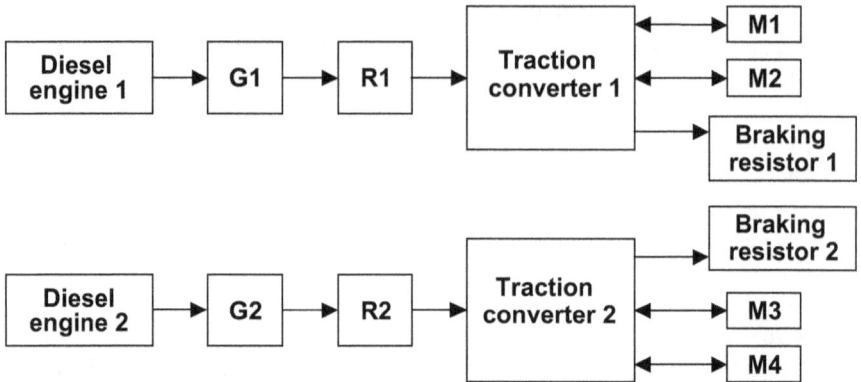

Figure 3.2 **Block diagram. Power circuits of the dual mode train AGC in diesel mode.**

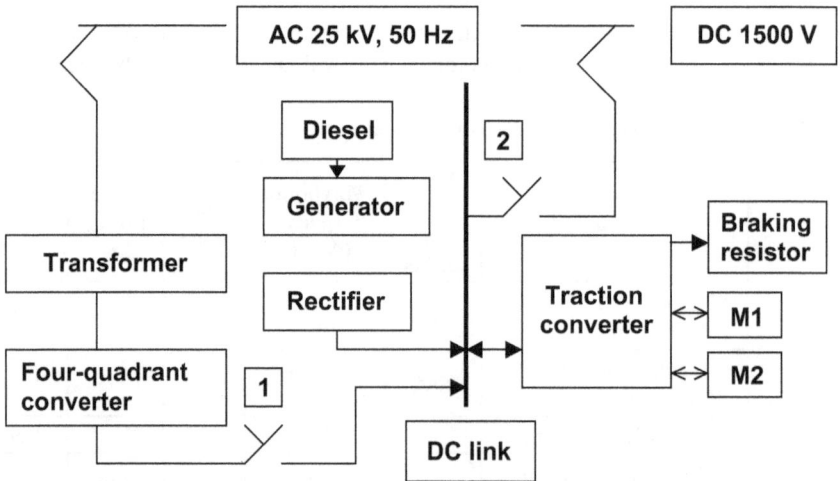

Figure 3.3 **Block diagram. Power circuits of the dual-mode and dual system train B 82500 (France)**

pantograph), also designated BGC; these trains are identical to the B 81500-series except for their capability of operating on AC power.

The block diagram of power circuits of the dual-mode B 81500 train is very similar to the block diagram of the tram-train Regio Citadis (Figure 3.1). The difference is only in the value of the DC voltage of the overhead line and DC link. In the case of B 81500, this value is 1.5 kV DC. The dual-mode and dual system train B 82500 can draw power not only from the overhead line of 1.5 kV DC but also from the overhead line 25 kV AC 50 Hz. The principle diagram of power circuits of the train B 82500 is shown in **Figure 3.3.**

The diagram shows only the power circuit and two motors. The main DC link operating at 1.5 kV DC has three possible sources of energy: in diesel mode it draws power from the diesel-generator set through the rectifier; in electric mode it draws power from the DC overhead line through the pantograph and switch 2 or from the AC overhead line 25 kV AC, 50Hz trough pantograph, transformer, four-quadrant converter, and switch 1. The switches are closed only if the train is under a suitable overhead line. The base characteristics of dual-mode/dual-system B 82500 series trains are very similar to the diesel train X 76500.

By August 2006 France Railways operated 135 X76500-series trains, 119 B 81500-series trains and 8 B 82500-series trains.

Placing transformers on the train is a challenge for both electric and dual-mode trains. The transformers operating at 15 kV, 16.7 Hz are large in size and heavy. Two transformers for the electric train Lirex weigh 12 tons. The transformer is generally installed beneath the car body. As a result, floors need to be raised and steps are required inside the train. The transformer contributes significantly to energy consumption during operations due to its weight and relatively high energy losses. Electric trains have two transformers, while the dual-mode trains have only one. The designers of the Lirex trains designed a future-proof approach (with lighter transformer in mind) to mount the electric power supply and drive components on the coach roof instead of mounting them below the body of the coach. Today, this is possible. The railway technology company, Alstom has developed a power supply system for electric trains that is up to 50% lighter and smaller than conventional systems. [3.8, 3.9]

The new technology offers considerable energy and cost savings and is seen as having a major impact on the design of future train generations. Because the newly developed system utilizes modern lightweight electronics in place of heavy iron and copper, these new systems have received the name of the "eTransformer."

The principle block diagram of power circuits of the dual-mode train Lirex is shown in **Figure 3.4.**

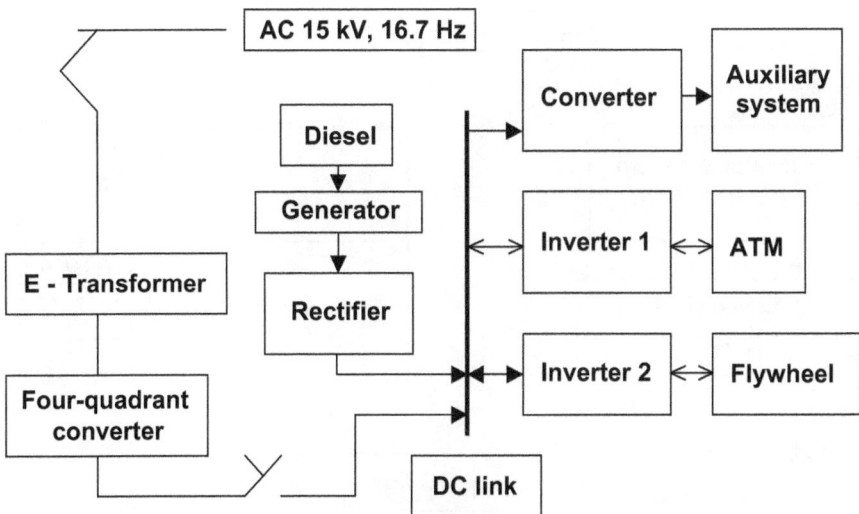

Figure 3.4 Block diagram. Power circuits - dual-mode train - Lirex (Germany)

In the diesel mode switch 1 is OFF. The diesel generator set feeds the traction motors through the rectifier and PWM inverter 1. When the train draws power from the overhead line, the switch 1 is ON and the traction motors are fed from the overhead line through the pantograph, "eTransformer," four quadrant converter and inverter 1. The technology is characterized by the use of a 4-quadrant controlled rectifier as a line-side converter in order to reduce stress on the power supply network and to ensure availability of a constant DC-link voltage and by the use of a pulse-width modulated (PWM) inverter as a machine-side converter for generating three-phase AC power.

3.3 Diesel and dual-mode locomotives
The majority of the modern freight diesel locomotives are diesel-electric locomotives. Every diesel-electric locomotive is ready to become a hybrid one. The diesel locomotives with asynchronous traction drives are closer to hybrids than diesel locomotives with DC drives. These locomotives have, as a rule, good efficiency, converters, and electric braking system. These locomotives generally have a greater power and adhesion coefficient than the other locomotives [3.10-3.13].

The characteristics of the modern diesel locomotives with asynchronous traction motors are listed in Table 3. 3.

Models SD70MAC and AC4400 (see Table 3.3) could serve a good example of modern locomotives with asynchronous traction motors. General Electric's model AC4400 has a system with individual controls for each motor where only one converter feeds one motor. The General Motor's model SD70MAC has another system in which one converter feeds 3 motors. The haul characteristics of these locomotives are similar.

The main circuit block diagram of the SD70MAC locomotive is shown in **Figure 3.5.**

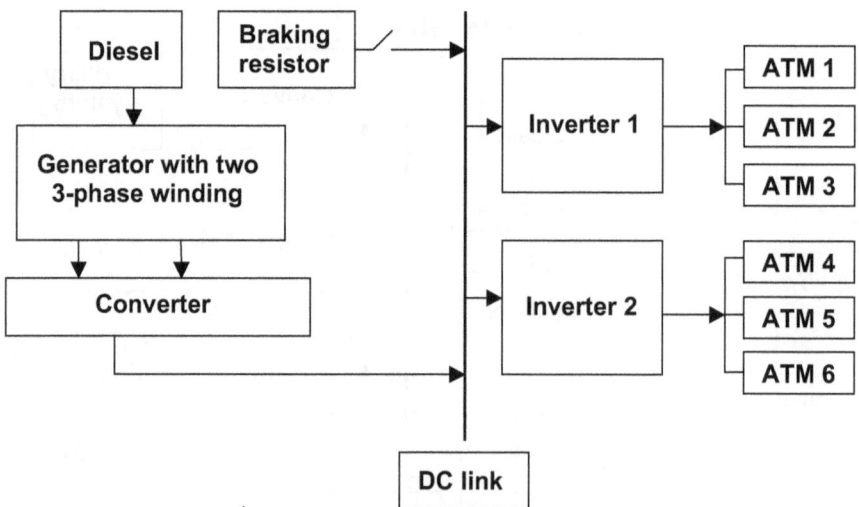

Figure 3.5 **Block diagram. Main circuit - SD70MAC locomotive**

Table 3.3 Main characteristics of modern diesel-electric locomotives with asynchronous traction motors

Parameters	Model of the diesel locomotives						
	SD 70 MAC	AC 4400	SD 80 MAC	SD 90 MAC	SD 70ACe	ES 44 AC	BB 475000
Country	USA	USA	USA	USA	USA	USA	France
Wheel arrangement	3o +3o	3o +3o	3o +3o	3o +3o	3o +3o	3o +3o	2o +2o
Weight	188/210	180	190	190	188	185	84
Length over couplers, mm	22555	22300	24400	24400	22936	22860	20280
Maximum speed, km/h	113	128	128	128	113	121	120
Diesel engine type	EMD 16-710 G3A	7FDL16	EMD 20-710 G3	EMD 16-710 G3	EMD 16-710 G3C-T2	GEVO 12	MTU 16V 4000 R41
Diesel engine rating, kW	2985/3150	3240	3730	4475	3160	3235	2000
Starting traction effort, kN	780	800	820	890	866	814	250
Continuous traction effort, kN	610	645	654	715	699	739	
Quantity of the motors, that are fed from one converter	3	1	3	3	3	1	4
Component of the converters	GTO	GTO	GTO	GTO	IGBT	IGBT	IGBT
Build date	1993	1994	1996	1998	2005	2005	2005
Number built	more than 1500	more than 2250	30	More than 100	about 1000	about 2000	400
Company name	EMD, Siemens	GE	EMD, Siemens	EMD, Siemens	EMD, Siemens	GE	Alstom, Siemens

- The converter includes two rectifiers.
- Each star winding feeds the rectifier.
- The power output terminals of two rectifiers have series connections.
- The DC voltage at the output of converter equals 2.6 kV DC
- Each traction inverter feeds three traction motors.
- The traction inverters are fed from the DC link.
- The switch closes the circuit of the braking resistor in the braking regime.
- Both the converter and the inverters use the diodes and GTO thyristors.

The GM/Siemens diesel locomotives SD80MAC and the SD90MAC have similar main circuit diagrams as model SD70MAC.

Models SD70ACe and ES44AC represent the new diesel locomotive line that meets the USA EPA's Tier 2 Locomotive Emissions Standards that took effect in 2005. The ES44AC model is one of the General Electric's Evolution Series locomotives. All the locomotives in this family are powered by the new GEVO 12-cylinder engine which delivers equivalent power as the old 16-cylinder engine, while consuming less fuel and producing fewer emissions. General Motors Electro Motive Division made the modification of the famous 16-710 engines that produces fewer emissions. IGBT's are used in converters for the SD70ACe and ES44AC which decreases the weight and size of the converters and increases their efficiency.

Model BB 475000 is a new diesel locomotive from the French National Railways SNCF. Model BB 475000 has superior characteristics compared to all the previous generations of European diesel locomotives. A consortium, consisting of Alstom Transport SA (consortium leader) and Siemens Transportation System, is planning to build four hundred locomotives of this kind. The electrical block, the diesel engine (MTU 16 V 4000 R41), the cooling system, the braking resistor, the control system and the cab display are already used in the RH 2016 diesel locomotive built by Siemens AG for Austrian Federal Railways (OBB). Siemens will manufacture one hundred locomotives for OBB. The main circuit diagram for locomotives 2016 and BB 475000 is shown in **Figure 3.6.**

Diesel engine (DM) turns the generator (G) in which the 3-phase voltage is rectified by the rectifier (R). The voltage from this rectifier provides traction PWM Inverter 1 to supply the four asynchronous three-phase traction motors and 1-phase Inverter 2 for the heating circuit of the train. The voltage from the exit of the rectifier also goes through Chopper 2 to supply the auxiliary services. The Inverter 3 generates a voltage/frequency controlled three-phase power to feed the main fan. Voltage and frequency are controlled in order to provide the necessary cooling output, thereby minimizing power consumption and fan noise. The second converter, (Inverter 4), is operated at a constant frequency of 60 Hz and a voltage of 440 V to feed the rest of the auxiliary services.

The energy from the regenerative braking is provided to the DC-link and is also used to feed the auxiliary services. If the amount of energy from the regenerative braking is more than can be used by the auxiliary services, the surplus energy is provided through Chopper 1 and to the braking resistor and is finally transformed to heat.

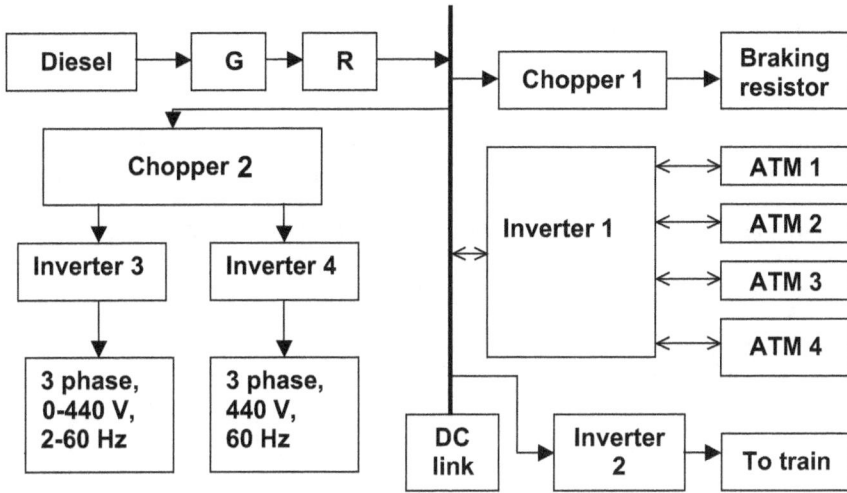

Figure 3.6 Main circuit diagram of locomotives 2016 and 475000

Dual-system locomotives:
• can take power from two different powered voltage lines
• can operate on both powered and not powered lines
Dual-mode locomotives became a combination of the diesel electric and pure electric locomotives. In the past, the diesel and electric locomotives were quite different. Today the difference between two has diminished. Modern diesel and electric locomotives use the same asynchronous traction motors, converters, and sometimes even the same mechanical parts. The main circuit diagram of the diesel locomotive BB 475000 **(Figure 2.6)** is very similar to the diagram of the modern electric locomotive. The comparison of two main circuit diagrams shows us that DC-link, the main and auxiliary services converters, braking circuits and traction motors are the same.

An example of the modern dual-mode locomotive [3.14-3.17] is listed in **Table 3.4.**
Dual-mode locomotives – main advantages:
• flexibility – operating on both electrified and non-electrified lines
• locomotive fleet reduction
• personnel training costs reduction
• maintenance costs reduction
The main circuit diagram of the dual-mode Siemens locomotive model 38 (South Africa) is shown in **Figure 3.7.** The first converter in the 3 kV DC circuit is the step-down chopper. The traction converter includes the inverter which feeds four traction motors and the chopper which connects with the braking resistance. When switch 1 is turned ON the locomotive runs in diesel mode. When that switch is turned OFF the locomotive runs in electric mode.

The main circuit diagram of the dual-mode locomotive model DM30AC(USA) is shown in **Figure 3.8.** The first converter in the 650 V DC circuit is the step-up chopper. In diesel mode, the DC voltage runs to the DC link from the diesel-generator

Table 3.4 Main characteristics of the modern dual-mode locomotives

Class of locomotive	ED1600	38	DM 30 AC	CC3600	ALP-45DP
Country	Germany	South Africa	USA	Spain	USA
Year(s) in service	1992	1992-1994	1997-1998	2007	2010-2012
Axle arrangement	2o +2o	2o +2o	2o +2o	3o+3o	2o +2o
Weight, т	88	74	128	130	131
Power in the diesel conditions, kW	425	780	2237	2*1800	3100
Line voltage	15kV, 16, 67 Hz	DC 3 kV	DC 650V	DC 3 kV	25kV, 60Hz 12.5kV,60Hz, 12kV,25Hz
Power by the feeding from electric net, kW	1600	1500	2150	4450	4000
The ratio of the power in diesel mode to the power by feeding from electric net	0,265	0,52	1,04	0.8	0.775
Track gouge, mm	1435	1067	1435	1435 or1668	1435
Starting traction effort, kN	360	260	360	445	316
Maximum speed, km/h	80	100	161	120	200/160
Numbers built	2	50	23	9	26
Company	Siemens	Siemens	Siemens	CAF	Bombardier

set and the rectifier. The switch system allows every traction converter to be turned off and uses the traction converter 2 for the feeding of the head-end power train line if the traction converter 3 is damaged. Contra versa traction converter 3 can feed the traction motors 3 and 4 if traction converter 2 is damaged. The braking resistor is activated only in braking mode.

The dual-mode models 38, DM30AC and ED 1600 use GTO-based converters. Newer dual-mode models used IGBT-converters.

New dual-mode locomotives of CAF (Spain) can operate on routes wired at 3 kV DC or on non-electrified lines. Versions can be built to run on 1435 mm and 1668 mm gauge with either 20+20 or 30+30 configuration. In February 2007 CAF received

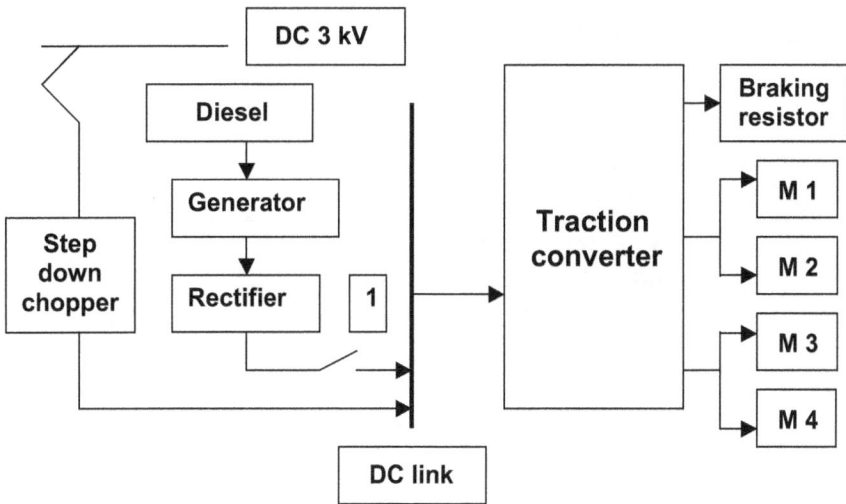

Figure 3.7 Main circuit diagram of the dual-mode locomotive model 38 (South Africa)

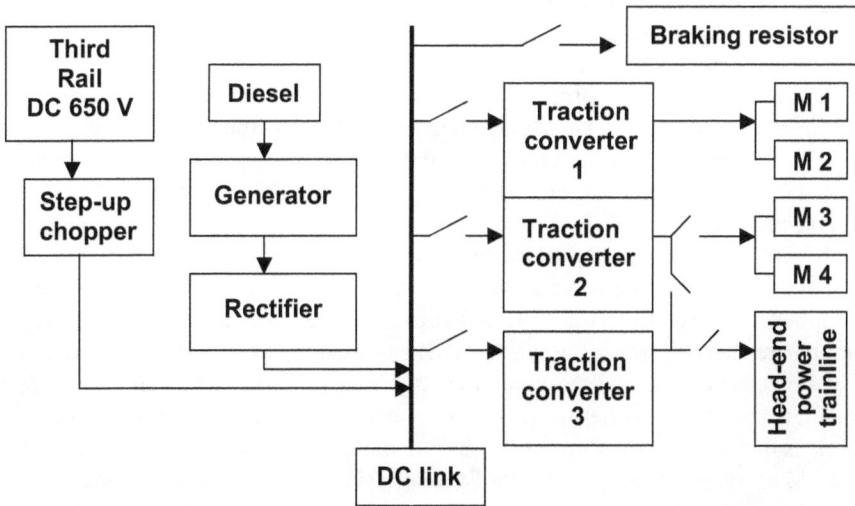

Figure 3.8 Main circuit diagram of the dual-mode locomotive model DM 30 AC (USA)

the order for 9 six-axle CC3600 freight locomotives. The characteristics of the locomotive are shown in **Table 3.4.** Locomotive with two diesel engine and IGBT-based converters has the power rating considerable higher than old locomotives.

The locomotive has two diesel engines, two main generators, two rectifiers, six IGBT-based inverters, six asynchronous traction motors and two IGBT –based 115 kVA auxiliary converters. The main circuit diagram consists of two identical schemes. Each scheme includes the power circuit of three traction motors.

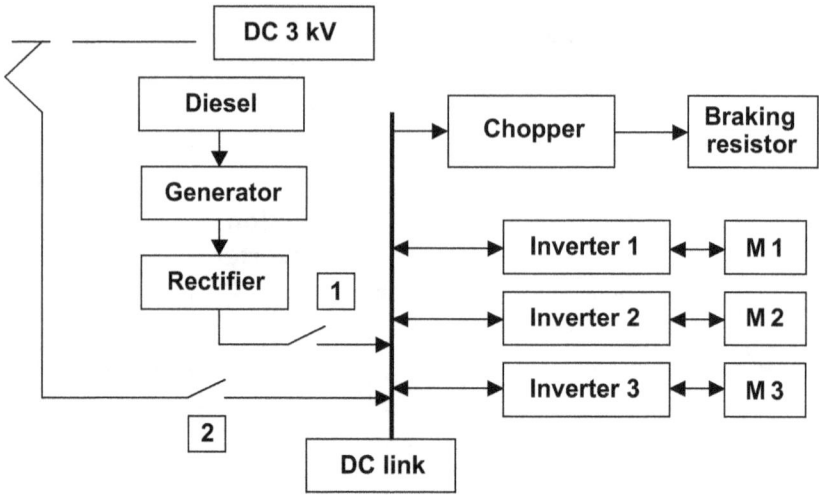

Figure 3.9 Main circuit diagram (half) of the dual-mode locomotive model CC3600 (Spain)

The main circuit diagram of the dual-mode locomotive model CC3600 (Spain) (single scheme) is shown in **Figure 3.9.**

Each Inverter feeds his traction motor. The braking resistor connects to the DC link through chopper only in braking mode.

When the switch 1 is turned OFF and the switch 2 is turned ON the locomotive runs in electric mode. The DC voltage runs to the DC link from DC wire net.

When the switch 1 is turned ON and the switch 2 is turned OFF the locomotive runs in diesel mode. The DC voltage runs to the DC link from the diesel-generator set and the rectifier.

Bombardier ALP-45DP dual mode passenger locomotive has very good prospects (3.17 - 3.18). The locomotive is based on the electric locomotive ALP-46A and is equipped with two Caterpillar 3516B diesel engines. The asynchronous traction motors are suspending on the tracks. IGBT transistors are used as converters. The diesel-electric engine is the source of energy in diesel mode. The main transformer and four-quadrant converters provide energy in the electric mode when the locomotive receives the energy through pantograph from the wire net. New Jersey Transit (USA) and AMT (Agence Metropolitane de Transport, Canada) will use the locomotives.

References

3.1. Siemens AG-Projects, Rolling Stock www.sts.siemens.com/p_nav4.html
3.2. Rob van der Bijl&Axel Kuhn "New Criteria for the "ideal tramtrain" Tramtrai1n. The 2nd generation. Workshop 2006 www.lightrail.nl/TramTrain/tramtrain.htm#Introduction
3.3 Regio Citadis. Information. Alstom LHB GmbH February 2004 http://www.kko.com/sprc/research/ConceptDualModeLightweightSPRC.pdf
3.4. Harry Hondius "Mixed fortunes in the railcar market" Railway Gazette International March 1999 P.147-152

3.5. Ali Asghar Shafi Naderi "Modern Diesel Trains". Rail International, 2003, #9, P.32-43

3.6. P.Villard, F.Ancelet "Automoteurs de Grande Capacite AGC" Revue Generale des Chemins de Fer, 2004, #7, P.7-24.

3.7. D.Lenhard et al. "Elektrische Ausrustung des triebzuges LIREX- Baureihe 618/619 fur DB Regio "Elektrische Bahnen. # 8, 2000, s.279-287.

3.8. "Electronic replace iron". IRJ, April 2004

3.9. " Electronic power supply system reduces operating cost". Innotrans 2004 Report.

3.10. Diesel-Electric Locomotive SD70MAC. Siemens Technical Information.

3.11. Rudolf Wagner "Antriebstechnik fur Diesellokomotiven in Nordamerika. ZEV+DET Glasers Annalen, Heft 2/3, 1994.

GE Transportation. Modern diesel locomotives.

Marl et al." The diesel locomotive Hercules with electric drive" Glasers Annalen, 2001, #6/7, S. 213-222.

3.14. Dual mode locomotives Class 38 and ED1600. Siemens Technical Information.

3.15. Diesel-Electric Passenger Locomotives DE30AC and DM30AC with Three-Phase Propulsion Technology. Siemens Technical Information.

3.16. Feve tests dual-mode loco. IRJ, December 2002. World Report

3.17. Electro-diesel loco flexible traction. Railway Gazette News 03 January 2008.

3.18. William C. Vantuono "NJT, AMT ready for dual-power locomotives", Railway Age, August 2008

4. DC TRACTION UNITS WITH ON-BOARD ENERGY STORAGE.

4.1. Light Rail Vehicle (LRV) with on-board ultracapacitor energy storage (Mannheim, Germany).

LRV is a rail borne vehicle lighter than a train, designed for the transport of passengers (and/or, very occasionally, freight) within, close to villages, towns and cities. Many newer light rail systems share features with trams, although a distinction is usually drawn between them, especially if the line has significant off-street running. The modern light rail systems are modular and fully flexible in vehicle length, width, number of cars, high or low floor and more. As a result of their flexibility, they have reduced delivery times and lower operating costs.

Since September 5, 2003, MVV Verkehr AG Mannheim (Germany) has been using the Bombardier Mitrac Energy Saver in daily revenue services [4.1-4.2]

A LRV equipped with the new Mitrac Energy Saver has proven 100 percent reliable in daily operations. The system works by charging up the energy saver during braking, and discharging it when the vehicle starts again to provide the necessary power for acceleration. The regenerative braking is practiced in conventional Mannheim LRV without the energy storage, but there have to be other trains around to absorb the surplus power being fed back to the overhead line. Too often, power produced by traction motors in braking mode ends up heating resistor banks. The alternative is to store the braking energy on the trains themselves.

The Mitrac Energy Saver allows the realization of this alternative. The Mitrac Energy Saver is a compact unit based on ultracapacitors that fits on the roof of a low-floor light rail vehicle. This LRV has two powered bogies, each with two motors. A Mitrac unit was connected in parallel to the DC link of the traction inverter producing variable voltage and AC frequency for the two motors of one bogie. The other bogie is supplied conventionally so that the energy performance of the two halves of the LRV could be compared. In addition to the ultracapacitors, the Mitrac unit contains a bi-directional step-down IGBT chopper which regulates the charge and discharge of the capacitors according to the requirements to store or export energy. The Mitrac unit weights about 450 kg, and the external dimensions are 1900 mm long, 950 mm wide and 455 mm deep. The energy of the one Mitrac Energy Saver is 860 Wh.

LVR has 4 traction motors, with power capacity of 95.5 kW. The length of LVR is 42.78 m and the width is 2.4 m. The track width is 1.00 m. The net weight of LVR is 48.65 ton. The maximal speed is 80 km/h; maximal deceleration is 3.0 m/c^2.

The advantages of the Mitrac Energy Saver integration are:
- Up to 30 % in energy saving
- Up to 50% reduction in peak power required from the network
- Up to 50% reduction in voltage drops via contact wires
- Reduction of direct energy costs, which are an increasing contributor to total life-cycle costs
- Possible saving in infrastructure-decreased line cross section, increased distance between substation

• Substantially decreases percentage of wasted energy during regenerative braking

LRV with Mitrac Energy Saver is able to accelerate rapidly for up to 1000 meters with a lowered pantograph by using the additional energy source. This means quite a bit of independency from the conventional power supply system. During a blackout, for instance, the next stop can still always be reached, and patches with the overhead line can be bridged thanks to the stored energy. Overhead lines can be even eliminated in front of historical buildings and monuments.

The successful tests of the trams with Mitrac Energy Saver in Mannheim allowed Rein-Neckar-Transport Network to make an order for 19 trams with ultracapacitor systems in 2007. Eight of them will be used in Heidelberg, the other 11 will run in Mannheim and on the railway-triangle between Mannheim, Weinheim and Heidelberg.

4.2. Tram Citadis with on-board battery energy storage (France)

Environment-friendly design is not just about noise, pollution and sustainable development. It also means full visual harmony with the city panorama. Some cities with historic architecture require alternative transportation solution. Overhead trams could not be deployed either in just the city center or across the whole system. To overcome these limitations and also to save additional energy, Alstom [4.2] offers its technology that allows to recover braking energy while remaining a zero-emission vehicle. Depending on the length of the section without overhead electrification, different technologies can be used.

For sections with limited length, such as a historic square in a city centre, high-speed performance is irrelevant. As an example, Nice in south-east France has two historic squares at the Place Massena and Place Garibaldi that required complete replacement of overhead wire for a distance of about 500m each. To meet this needs, the city chose Citadis trams. These 20 vehicles, to be delivered in 2007, will carry NIMH batteries, which offer a good compromise between performance, weight, volume and life-cycle cost, allowing the trams to cross the two squares without overhead wire at a lower speed. The system will be compact enough for installation on the vehicle's roof. The 576V Saft NiMH battery system provides 80 kWh of continuous power. Each battery system incorporates an active cooling device and battery management control (BMC) for consistent monitoring of temperature, voltage and charging conditions.

4.3 Tram Citadis with on-board flywheel energy storage (France)

The concept of flywheel storage offers a new promising way of storing energy on-board. It allows trams [4.2] to operate without overhead wire sections and to recover braking energy, thus saving energy and permitting system optimization. The advantages of the flywheel energy storage are better life cycle, power density as well as rapid charge rate and storage efficiency. Modern flywheels are made of composites with integrated electrical equipment and enclosed in a safe vacuum container. Flywheels are designed to rotate at speeds above 20,000 rpm, providing net energy storage of 4 kWh and up to 325 kW during peak loads.

The flywheel of tram in Rotterdam has the dimensions: length – 6.56 ft; width-4.59 ft; height-1.67 ft. The masse of flywheel is few more than 1 ton.

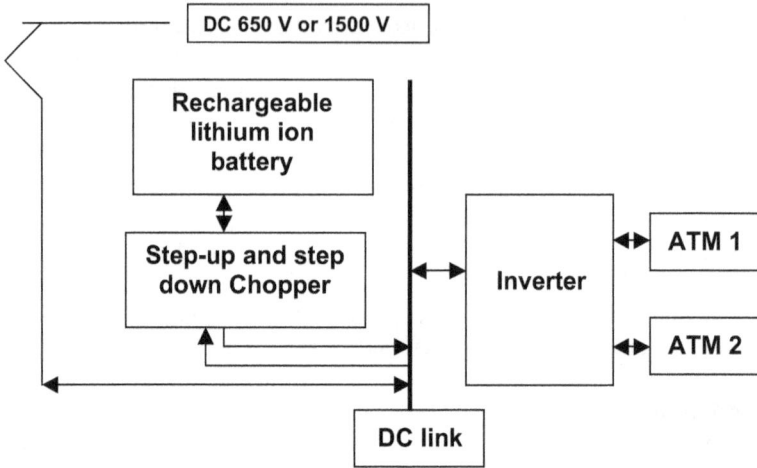

Figure 4.1 Block diagram of the energy system for the energy recycling tram and EMU with Li-Ion battery energy storage (Japan)

Not so big dimensions and weight of flywheel system allows mounting it on the tram roof.

The flywheel stores the energy generated during braking and makes it available when the vehicle accelerates to offer two advantages:

• For sections without overhead wire, the vehicle can run at a reasonable speed with good acceleration. Thanks to its rapid charge rate, the flywheel can be "replenished" while the vehicle stops, allowing operation in a city free of overhead wires.

• For sections with an overhead wire, the flywheel provides load leveling to reduce investment in ground infrastructure.

The test of tram Citadis with flywheel in Rotterdam at 2005 shown that this tram spent 15% energy in comparison with common tram without flywheel. The life cycle of flywheel system is 30 years. Same as tram themselves.

4.4. Tram with Li-Ion battery energy storage (Japan)

On February 2, 2005 the Railway Technical Research Institute (RTRI) of Japan announced that it had successfully completed a test run of its tram, which uses both on-board batteries and electricity from overhead lines. [4.3-4.5]

Several trams have a regenerative braking capability, which means that when a tram is braking it generates electricity which is fed back into the overhead lines, where it can be collected and used by other trams in the same area.

However, regenerative braking loses its effectiveness when there are no other trains running nearby that can immediately use the regenerated electricity. In such a case, the train must use friction brakes, which do not convert kinetic energy into electricity and also are subject to wear and deterioration of their parts. To cope with these disadvantages, the structure of the new tram allows the electricity regenerated during braking to be stored in the batteries. The stored electricity is also used as energy for restarting the tram, which reduces the amount of power needed from overhead wires.

The RTRI tested the powering and regenerative braking performance of the tram cars at the upper limit of contact wire voltage, the behavior when the pantograph is reduced or raised, and other response characteristics in abnormal situations. Results demonstrated that electric energy that could not be returned to the contact wire was stored in the onboard charging battery and that tram cars accelerated or decelerated smoothly (even in abnormal situations) with power supplied by the contact wire or from the charging battery. In the running test in the compound, the brake energy recovery rate was about 70% when brakes were applied at 50km/h.

A block diagram of the energy recycling tram is shown in **Figure 4.1.**

The tram system is fed from the DC wire net at 650 V and consists of an inverter, induction motors, a current reversible converter, and a rechargeable lithium ion battery. The traction controller and energy storage equipment are configured in parallel. In this system, braking energy charges up the lithium ion battery and is regenerated to the power feeder line. The battery also supplies a part of the energy required when the tram is accelerating. Therefore, the storage system using the battery can avoid regenerative braking force cancellation and achieve a high regenerative ratio.

The length of the tram is 11500 mm; the width is 2200 mm, new wheel diameter is 720 mm. The empty vehicle weights 19700 kg.

The electric equipment of the tram includes a rechargeable lithium ion battery, Pulse Width Modulated (PWM) inverter (the power of inverter is 180 kW) and two traction induction motors. The power from each motor is 60 kW.

The acceleration of the tram is 3.0 km/h/s (up to 29 km/h), the deceleration is 3.4 km/h/s.

The capacity of the battery in regime C is 55 Ah, the voltage is 604.8 V (3.6 V*168 cells). The maximum current of the battery is 500 A. The weight of the battery is 1160 kg (two blocks, each block includes 84 cells).

4.5. Electric Multiple-Unit Train with on-board Li-Ion battery energy storage (Japan)

In the 1980s, asynchronous traction motors replaced direct current motors for almost all EMUs of Japan National Railways and regenerative braking capability became inherent in the power control circuits. The problem of EMUs with the regenerative braking was an occasional regeneration failure due to the lack of line receptivity. RTRI tried to find a solution to this problem [4.3-4.5]. The solution relied on storing regenerative energy on board of the EMUs.

RTRI used a lithium ion battery and developed a power converter (a step-up/step-down chopper) that can be installed on existing EMUs, and an algorithm to control it. A block diagram of the energy recycling system is shown in **Figure 4.1.**

This has enabled EMUs to control power supplied through contact wires and from the on-board battery. As a result, the power generated in braking is stored in the on-board battery when it cannot be returned to the contact wire. This solution substantially decreases percentage of wasted energy during regenerative braking.

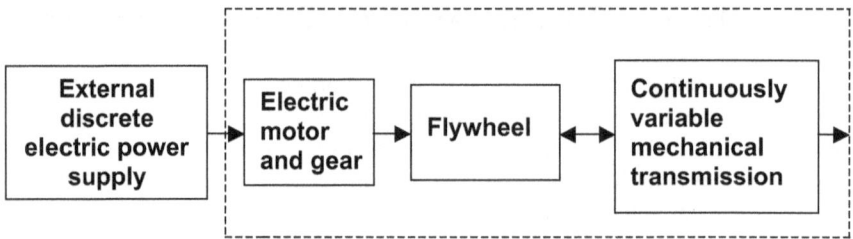

Figure 4.2 Block diagram of the PPM 35 vehicle

The RTRI tested a 1,500V EMU with the speeds up to 100km/h at its rolling stock test plant. The desire was to check its behavior at shutdown and recovery of contact wire voltage in powering and regenerative braking under hybrid control.

There will be fewer cancellations or breakdowns of the regenerative brakes when EMUs with on-board energy storage are installed as opposed to the common DC EMU. Damage to the energy storage is rare and regenerative braking with ES is safer than former systems.

The effective utilization of regenerated power to save energy, improve the reliability of the regenerative brake and reduce activation of the mechanical brake are expected to lead to labor savings in the maintenance of brake shoes and wheels.

The EMU set includes 4 motor cars and 6 trailers. The tare weight is 256.4t. Acceleration and average deceleration are equal to 3.0 km/h/s.

4.6 Railcar Parry People Mover PPM 35 (England)

Parry People Mover Ltd was formed in 1992. The task of the company was to develop a hybrid vehicle using energy storage. The company uses the flywheel as its main energy storage source. The flywheel of PPM is simple, reliable and easily maintainable. It is made from steel and has a 1m diameter and 500 kg mass. The maximum speed of the flywheel is 2500 rpm [4.6]. A feature of the PPM vehicle is a lack of the complete electric drive. PPM uses the components of the mechanical and hydrostatic transmissions. A block diagram of the PPM 35 vehicle is shown in **Figure 4.2.** As in other hybrid rail vehicles, the PPM 35 has the ability to use the braking energy and the ability to store the kinetic or potential energy during the brake time and feed it back during acceleration or during maximal power demand. In the case of PPM 35 vehicles, the flywheel can be charged approximately for 30 seconds from an intermittent electrical supply - at the station only. The electric motor untwists the flywheel through the gear; the energy of the flywheel goes through continuously variable mechanical transmissions to the wheels. Between 1998 and 2000, a PPM 35 railcar carried over 50,000 passengers in Bristol (England).

References
4.1. Richard Hope" UltraCaps win out in energy storage" Railway Gazette International July 01 2006.
http://www.railwaygazette.com/Articles/2006/07/01/2776/UltraCaps+win+out+in+energy+storage.html

4.2. Francois Lacote "Alstom-Future Trends in railway Transportation". Japan Railway &Transport Review 42, December 2005, p. 4-9

4.3.Hiroshi Sameshima, Masamishi Ogasa, Takamitsu Yamamoto " On-board Characteristics of Rechargeable Lithium Ion Battery for Improving Energy Regenerative Efficiency" QR of RTRI, Vol.45, #2, May 2004.

4.4. M.Ogasa, H.Sameshima. "Tram with batteries power supply". Japanese Railway Engineering, # 1, 2004, p.23-26.

4.5. Hybrid EMUs mounted with charging battery. Main Research and Development Results for fiscal 2004> 4 Harmonization with the environment.

http://www.rtri.or.jp/rd/openpublic/seika/2004/04/environment_E01.html

4.6 Parry People Movers- Technology 16 January, 2006

http://www.parrypeoplemovers.com/technology.htm

5. EXAMPLES OF THE REAL HYBRID TRAMS, HYBRID DIESEL TRAINS AND HYBRID LOCOMOTIVES.

5.1. The world's first hybrid railcar and the hybrid diesel-battery multiply unit (DMU) Kiha E200 (Japan)

Japan Rail East, Hitachi, and Tokyo Car Corporation together developed the New Energy Train, or NE train for short [5.1-5.4]. The first step of the project was to install hybrid traction systems consisting of an engine, generator and storage battery in a Class E991 diesel railcar. Testing started in May 2003. The goals of the collaboration were the improvement of diesel car energy efficiency, as well as reduction of the noxious emissions and noise. This was the first train car in the world to use a hybrid system.

The block diagram of the NE train is shown in **Figure 5.1.**

The series hybrid system converts the engine output into electrical power and then uses only motors for propulsion. The main converter includes both the converter and inverter. The AC (alternating current) output generated by the engine is converted into a VVVF (variable voltage variable frequency) AC supply by the main converter to drive the induction motors.

Storage batteries are attached to the DC link of the main converter, and the charging and discharging of the storage batteries are controlled using output adjustments of both the converter and inverter.

This series hybrid traction system allows the engine speed to be set irrespective of the vehicle speed. This decision allows the receipt of high-efficiency power generation and reduces exhaust gases. Most of the time the engine works in the speed range of low fuel consumption.

The inverter that is used in the power circuits allows the ability to invert the energy of the regenerative braking into charging the accumulator's battery and then use this energy for the next acceleration. The use of the regenerative energy allows the reduction of fuel consumption approximately to 20% compared with conventional diesel trains.

The basic principle of the Energy Management and Control System (EMCS) is to control power generation of the engine in such a manner that "the sum of energy generated by the car motion (which changes with car speed) and stored energy (stored in the battery) is kept constant regardless of speed." [5.1]

The State of Battery Charge (SOC) is sustained in all regimes between 20% and 60%.

The work of the traction drive is divided into four areas depending on the speed and SOC.

In the regime of departure from station the storage batteries alone are used for acceleration at low speeds. In the mid-speed range additional power is provided by the diesel-generator. During cruising, the diesel-generator works in the optimal regime. The battery charges or discharges depending on running loads. While running up a slope, the diesel-generator works with full power and the battery helps. While running down a slope the battery is charged by the regenerative energy. It also uses the engine exhaust brake. When braking, the engine is shut down and regenerated

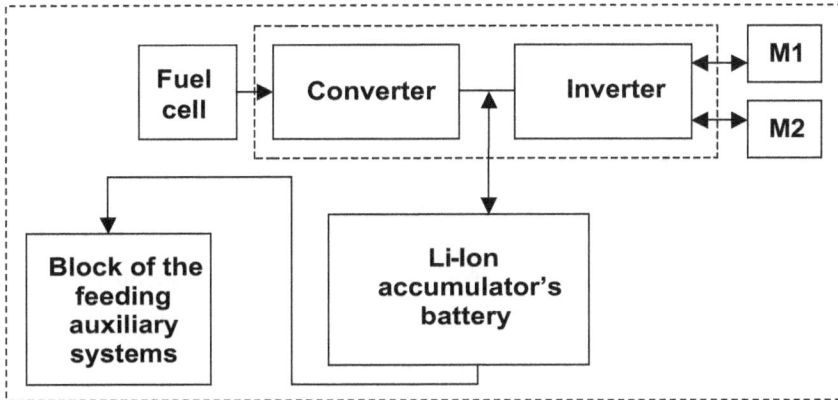

Figure 5.1 Diagram of the series hybrid system of the world's first hybrid diesel train (Japan)

braking power is stored in the batteries. When making a stop at a station the engine is shut down to reduce noise in the station and to improve fuel consumption. Service energy is supplied from the battery.

The hybrid propulsion system stabilizes the speed on downhill gradients with the help of the engine brake control. The braking power is generated by the traction motors and inverter to drive the converter and generator in reverse and absorb energy with the engine load. The advantage of this decision is the independence of the engine braking power from the vehicle speed and the possibility to provide the speed stabilization under any running condition.

After traction calculations and economical estimations the following parameters were selected for the main devices of the hybrid system:
• The power of each traction motor is 95 kW
• The power of a newly developed low-emission diesel engine is 330 kW
• The power of a induction generator is 180 kW
• The energy of a lithium-ion battery is 10 kWh and the output is 250 kW
Electrical components and the driver's cab of the NE train are similar to those used in Tokyo commuter E231 EMUs.

In the running tests conducted on the Nikko, Karasuyama, and Tohoku lines, the following performance items were reviewed:
• Technical maximum speed is 100 km/h
• Acceleration performance is 0.64m/c² (at 35 km/h)
• Deceleration performance is higher than 1m/c²
• The car was able to accelerate up to 70 km/h with the use of the battery energy (travel distance: 2 to 3 km)
• The car could run with only an engine when the battery was cut off due to failures.
• The engine stopped (for more than 5 minutes, although it depended on the air-conditioning load) when the car stopped at a station, and the engine remained off when the car started again (until it reached approximately 25 Km/h).
• Car performance was examined under outside air temperatures of 35 C, as well as under sub-zero temperatures and no abnormalities were found.

• The target value of reduction of energy consumption by 20% was close to being met.
• The rate of energy regeneration was approximately 20%.

The successful trial of hybrid railcar gave the East Japan Railway the possibility to begin commercial service of hybrid railcars. On July 31st 2007 East Japan Railway put it's first diesel-electric Kiha E200 hybrid train into commercial service. The train includes two hybrid railcars. These railcars together have 46 seats and can hold 117 people including standees. The train is debuting on a mountainous line which has a length of 79 km (49 miles).

This project has boosted fuel efficiency by 20 percent and reduced exhaust emissions by up 60 percent. The price of hybrid train ($ 1.7 million) is double the price of the conventional train.

The commercial service of three Kiha E200 trains which were launched on the Koumi line in July 2007 was successful. In autumn 2010 East Japan Railways will introduce 10 diesel-electric hybrid multiple units.

5.2 Hybrid regional innovative train LIREX (Germany)

Light Innovative Regional Express or LIREX is the light articulated train that is aimed at medium-distance services. [5.5 – 5.9]

The LIREX was developed by Alstom together with Deutsche Bahn AG (DBAG) and promoted by the Land Saxony-Anhalt, Germany. It is designed to be an innovative test train for concepts and components.

This new train can be powered by either diesel or electric traction, or both. The prototype is a dual-mode electro-diesel. When building the train it was required to have an ability of the regenerative braking (diesel train mode included).

The features of the LIREX are:

• The articulated train is mounted on single-axle wheelsets, which reduce the weight of the running gear by half compared with a train with conventional bogies.
• The train has a low floor throughout its length.
• Most equipment is mounted on the roof.
• The train has flywheel storage, which makes it possible to store energy during diesel-electric operation, which can then be used to power the train.
• The train is composed of three basic modules. There is a so-called centre car (M) mounted on two single-axle wheelsets. The other two car types (A and B) have only one single-axle wheelset per car. The A car has a cab. In the electric only train, the A and B cars have traction equipment. In the diesel train, every car has traction equipment. The basic train has six cars. Additional B and M cars can be added to form longer trains.

The key technical data of the Lirex train are summarized in **Table 5.1**.

A block diagram of the power circuits of a 3-car section is shown in **Figure 5.2**.

Two diesel-generators are connected through the rectifiers with a DC-link. Four converters are fed from the DC-link. A pulse-width modulated (PWM) converter is used as a machine-side converter. All 3 traction motors are connected in parallel. Three of the four axles of the 3-car section are powered. The three-phase auxiliary converter and converter of the flywheel also are fed from the DC-link. Every converter uses IGBT

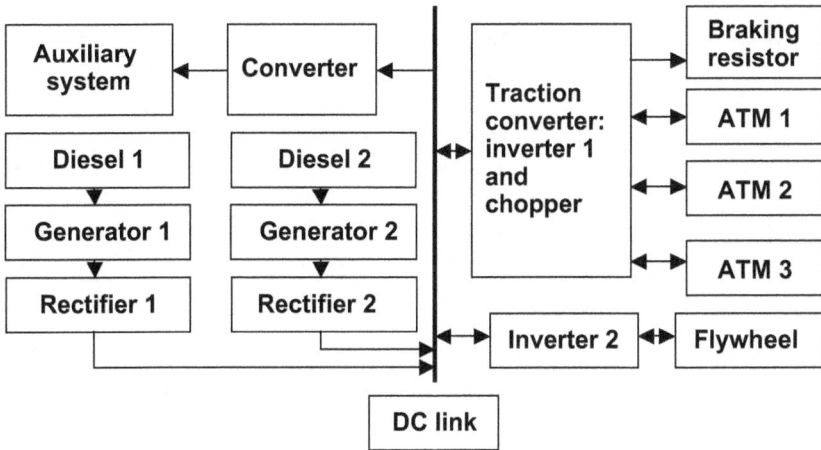

Figure 5.2 Diagram. The power circuits of the Lirex train in diesel mode.

Table 5.1. Main technical characteristics of the Lirex train

The number of car	6
Length over couplers, mm	68490
Carbody width, mm	3042
Height of roof profile, mm	4500
Service weight, t	137
Number and power of the diesel, kW	4*338
Number and capacity of fuel tank, l	4*800
Number and power of the traction motors, kW	4*190
Maximum starting effort, kN	120
Maximum speed, km/h	160

transistors. Electric braking energy is transferred to the flywheel. The surplus energy transfers to a brake resistor. The brake resistor can take all the power that is generated by the braking in a middle speeds range. The train also has disc and magnet-rail brakes

Four diesel-generator sets are roof-mounted. The six-cylinder 338 kW diesel engine is a well-proven MAN unit. The generator is a synchronous three-phase electric machine with excitation from permanent magnets. Its masse and size are considerably less than that of conventional generators. The rectifier is mounted directly on the frame of the generator's stator.

The six 190 kW asynchronous water-cooled traction motors are mounted on the wheelsets. The traction motors are based on an existing design but have a higher output.

Lirex is the first in the world to have multiple diesel units with a flywheel energy storage system [5.7]. The flywheel accumulates the energy during the braking. The accumulated energy can be used for the following:
• Increase the power during the acceleration
• Reduce the fuel consumption by the reduction of the full load of the diesel engine
• Proceed in several short stages without a noise and without exhausts.
High-speed kinetic storage operates at speeds up to 20,000 revolutions per minute. The mass moment of inertia, weights and dimensions are relatively small. The flywheel stores energy while the motor/generator unit is run as a motor, accelerating

Table 5.2 Technical data of the Lirex flywheel

Energy content	6 kWh
Maximum power	350kW
Efficiency including frequency converter (charging/recharging)	> 90%
Idling losses	2.5-7 kW
Voltage, V	550-750
Rotor material	Carbon fiber/epoxy resin
Diameter of the rotor	700 mm
Maximum speed	25.000 r/min
Minimum speed	12.500 r/min
Type of bearing	Precision ball bearing with lubricating oil circuit
Type of motor	Synchronous motor, permanent excitation
Life	20 years
Suspension of storage flywheel	Resilient mounting
Working temperature range	From -40 C to 60 C
Dimension of the complete system	1900*1625*1080 mm3

the rotor. The flywheel feeds back energy when the motor/generator unit is switched to generator mode, thus reducing the rotor speed. Technical data of the Lirex flywheel are shown in **Table 5.2.**

The Research Center of Deutsche Bahn AG carried out a simulation of the operation of the Lirex with energy storage, showing an energy saving potential of about 11% for a vehicle with flywheel storage compared with a similar vehicle without energy storage. The simulation was carried out for the German line having a total length of 87 km and 8 stations. The simulations showed that the use of storage flywheels in vehicles produced the greatest savings on routes with short distances between stops.

A block diagram of power circuits of 3-car sections of dual-mode Lirex train was already shown in **Figure 3.4**. It is not in as much detail as a block diagram of the diesel section in **Figure 5.2**. The three-phase auxiliary converter and converter of the flywheel are fed from the DC-link in the power circuits of the dual-mode Lirex train as well.

Alstom has used flywheel as an energy storage unit in different experimental trains including Lirex, Coradia Lirex, dual mode tram-train Regio Citadis(Figure 3.1) and Citadis tram (Figure 4.3, page 57).

5.3 Hybrid railcars PPM 50 and PPM 80 (England)

Hybrid railcars PPM 50 and PPM 80 are the next step of development of Parry People Movers Ltd. vehicles after PPM 35(see chapter 4.6).

The PPM 50 and PPM 80 as well as PPM 30 have the flywheel that stores the kinetic energy of the vehicles. PPM 50 is a hybrid rail vehicle, PPM 35 is a vehicle with energy storage but it is not a hybrid. The prime mover PPM 50 is the Ford 2 liter LPG engine. LPG (liquid petroleum gas) is widely used as a "green" fuel alternative for internal combustion engines as it significantly decreases exhaust emissions. LPG engine's trough transmission untwists the flywheel and the energy from the flywheel goes through hydrostatic transmission to the wheels. The block diagram of the PPM 50 vehicle is shown in **Figure 5.3**.

PPM 50 can transport 50 passengers. The flywheel has the same parameters as the flywheel of PPM 35 (see the part 3.6). PPM 50 is practically a zero-emission railcar. The scientists of the Parry People Movers Ltd. confirm [5.15] that 50 people can travel on one gallon of fuel in a common diesel railcar 2 miles, a common bus 7 miles and in the PPM 50 15 miles.

PPM 80 has a maximum capacity of 80 (Including standing passengers). It is 13.7 m long and has a weight (tare) of 14 tons. The maximal speed of PPM 80 is 50mph (80 km/h)[5.16]. In contrast to the PPM 50, the new railcar has two Ford 2 liter LPG engines (two powered bogies) and two flywheels.

PPM 80 is an ultra light railcar. PPM 80 can be used as a tram. The advantages of this tram are light weight and autonomous method of operation.

With using overhead continuous electrification, conventional trams require grounding the current through the rails. This can cause electrolytic damage and potential arcing if the segments aren't properly joined. Underground insulation and removal of utility services from under the rail path are the ways to avoid the

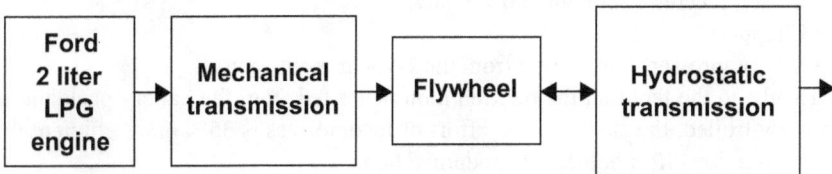

Figure 5.3 Block diagram of the PPM 50 hybrid vehicle

shortcomings of the return currents. But these solutions are expensive. PPM 80 trams don't need underground insulation and it is its big advantage. Buried services do not have to be removed and tracks can be easily relocated if required. The new hybrid rail vehicles of Parry People Mover Ltd. combine the minimal weight and energy storage. This has the advantages of [5.15-5.16]:
- Decrease in power and the size of the prime mover
- Small amount of material and energy consumed during manufacturing
- Lack of perceptible emissions
- Gain of energy in maintenance
- Possibility to use lightweight infrastructure that adds up to cost savings
- High service reliability and punctuality

5.4 Modernized Hybrid Switch Locomotives
5.4.1 Hybrid Switch Locomotives Green Goat (GG20B)
Canadian company Rail Power Technology modernized the traditional switcher diesel-electric locomotive and put into practice a new technology. [5.10-5.14]. This locomotive radically reduced fuel consumption and emissions of pollutants after modernization. The name of the new locomotive is Green Goat. The word Goat is slang for a locomotive called a Switcher Engine or a Yard Switcher. The word "Green" is a universal color usually associated with the earth and its environment. A diesel-generator set of locomotives is changed by a big accumulator battery and small diesel-generator sets during modernization. Traction motors of modernized locomotives are fed from an accumulator battery. A diesel-generator set of the new locomotive is used for the permanent charging of accumulator batteries. Thanks to the fact that it works with diesel-generator sets, it has a constant revolution per minute and the level of charge of the battery is always more than 80%. After modernization, the switcher locomotive works practically as an electric switcher with an autonomous diesel electric source of charge. The locomotive has the choppers for regulating traction motor's voltage.

The frame of an EMD GP-9 locomotive was used in constructing the Green Goat.

GP9 is a convectional General Electric EMD diesel switch locomotive of 119 t and 1750 HP, built circa 1958. The Green Goat also used the same GP-9's conventional 2100 US gallon fuel tank, existing traction motors and switchgear. All components, including the control stand, trucks, wiring, brake system, traction motors and switchgear, are remanufactured to the same specifications as a new locomotive.

New components include:
- Maintenance free absorbent glass mat (AGM) valve regulated lead acid battery
- Microprocessor control system with individual slip control
- Diesel-generator set with 200 kW power
- Choppers
- Air compressor with a drive from the asynchronous motor

Thanks to the fact that the traction motors are fed from the battery pack and the slip is controlled, the start traction effort of locomotives is 35% more efficient than the start traction effort before the modernization.

A block diagram of Green Goat power circuits is shown in **Figure 5.4**

The benefits of the modernized locomotive are:

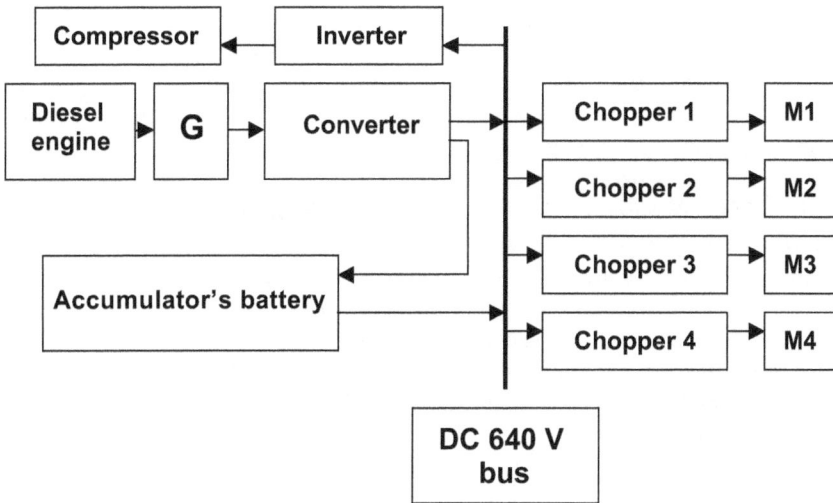

Figure 5.4 Block diagram of Green Goat power circuits

• Reduction of fuel consumption
• Reduction of emission polluters
• Reduction of the noise and the maintenance expenses

The Green Goat achieves a 90% emission Nitrogen Oxides (NOx) and Carbon Monoxide (CO) with a 77% reduction (from 19.5 grams/hp hour to 4.5 grams/hp hour) of Particulate Matter (PM) compared to typical 2000 horsepower locomotives.

The modernized locomotive has the wheel arrangement Bo-Bo.

The locomotive Class GG 20B has a diesel engine Caterpillar C9/Leroy Somers with an output of 200 kW, 480 VAC.

The Green Goat utilizes a rotary screw type air compressor driven by a 30 HP A/C motor. The air compressor's A/C motor is powered from the locomotive batteries through an inverter.

The Green Goat uses over 50,000 pounds of batteries. These batteries account for 95% of the Green Goat's horsepower. A total of 320 two-volt absorbent glass mat (AGM) valve regulated lead acid battery cells are used. Valve regulation provides better control of hydrogen gas release when the batteries expand at different temperatures. AGM batteries are constructed to be sealed, leak proof, and vibration resistant. They use a fiber floss glass mat with wicking characteristics tightly sandwiched between the lead plates, which allow them to be constantly saturated in electrolyte. This allows for a faster reaction between the acid and plate material.

These battery cells are divided into 40 steel racks containing 8 cells each at 16 volts per rack. Combined, they make a total of 640 volts and have a total capacity of 1200 ampere hours.

Voltage on groups of the battery cells is monitored by the on-board computer system, which provides instant notification of poor cell condition. If a battery cell does fail, it can be manually bypassed in the field until a scheduled repair can be made. Up to 10% (or 32) of the battery cells could be bypassed at one time. Ninety-two day scheduled maintenance of AGM batteries consists only of cleaning the battery cell racks and inspecting the battery terminals for corrosion. Since the AGM batteries are sealed (maintenance free), electrolyte is not checked. Battery recycling is recommended after a target ten years of service.

The Green Goat's batteries receive a constant trickle charge from a generator driven by a 200 kW diesel engine. The batteries are kept to a partial state of 80% charge, which increases the batteries' efficiency and extends their lives.

With the batteries producing 95% of the Green Goat's horsepower at fuel load, the other 5% comes from the diesel generator set, supplying the locomotive with traction horsepower when needed or trickle charging the batteries. The engine operates at a set constant speed that is tuned for maximum performance and efficiency. Since it uses standard diesel fuel, the Green Goat has the flexibility to utilize the existing railroad fueling infrastructures.

The specification of the locomotive GG20B is shown in **Table 5.3.** [5.11].

New 1500-2000 HP switchers cost 1-1.5 million dollars. (All economical data used is based on RailPower data.) A typical refurbished diesel switcher costs about $450,000. The remanufacturing of the traditional diesel switcher to Green Goat costs $750,000. A typical diesel switcher consumes 250 gallons of diesel fuel in a normal day. The Green Goat consumes 65% less, saving $140,000 a year. Moreover, typical switching locomotives cost $100,000 a year to maintain; the Green Goat's upkeep costs $20,000. Green Goat repays the remanufacturing costs of the traditional diesel switcher very fast. The capital cost of a hybrid locomotive would be recouped in less than three years. In addition to being cost-effective, Green Goat's application leads to significant reduction of GNG emissions and to the decrease of air contaminants such as nitrogen oxides and diesel particulates.

The design and trials of the Green Goat continued from 1999 to 2003.

At first the Company tested the locomotive on its own networks with its own staff. Then the Green Goat was leased to Union Pacific for lengthy trials. After nine months of testing at Roseville, California, it was moved to one of the company marshalling

Table 5.3. Main technical characteristics of hybrid switch locomotives GK10B, GG20B, and RP20BH.

Class of locomotive/Parameters	GK10B	GG20B	RP20BH
Axle arrangement	2o + 2o	2o + 2o	2o + 2o
Power of diesel engine, kW	90	200	2 diesel engine 500 kW each
Parameters of the storage battery	320V,1200A	640V.1200A	700V, 600AH
Traction horsepower equivalent, hp	1000	2000	2000
Starting traction effort, kN	363 to 8 km/h	408 to 8 km/h	454 to 20.5 km/h
Maximum Speed , km/h	48	96	104.6
Maximal Weight (full service), t	112.5	124.74	124.74

yards in Chicago so that its performance under different winter conditions could be evaluated.

RailPower Technologies Corp. reached a partnership agreement with United Goninan Ltd. to help the locomotive builder market its hybrid switchers in Australia and Southeast Asia. United Goninan designs, manufactures and maintains railway rolling stock at 14 facilities throughout Australia and Asia. RailPower Technologies Corp. also signed a memorandum of understanding with Mitsui Bussan Transportation System Co. Ltd. to explore the market potential for RailPower's hybrid locomotive technology in Japan, and with Swedish Train Technology («STT»), to explore the market potential for RailPower's hybrid locomotive technology in Scandinavia.

5.4.2 Hybrid Switch Locomotive Green Kid (GK 10B)

The Green Kid is a less powerful version of the Green Goat for those jobs where such massive power is not necessary. All principal decisions are the same.

While the 2000 HP Green Goat is designed essentially for use in large freight yards, a smaller, lower-power 1000 HP Green Kid was considered to be more suitable for work in industrial complexes, where movement of freight wagons from loading facilities to a junction with the main line, rather than the marshalling of trains, is the primary rail activity. The specification of the locomotive GK10B is shown in **Table 5.3.**

In constructing the Green Kid, the frame of a Class SW1200 locomotive was used. The Green Kid uses the SW1200's same conventional 600 US gallon fuel tank, existing traction motors and switchgear.

The switcher Green Kid has several differences from the switcher Green Goat:

The Green Kid uses over 25,000 pounds of batteries. These battery cells are divided into 20 steel racks containing 8 cells each at 16 volts per rack. Combined they make a total of 320 volts and have a total capacity of 1200 amp. hours.

The locomotive Class GK 10B has a diesel engine type called Duetz/Marathon with an output of 90 kW, 240 VAC.

The design and trials of the Green Kid were finished circa 2003-2004.

5.4.3 Hybrid Road Switcher RP20BH

RailPower designed its RP series specifically to reduce high fuel usage in road and branch line switching operations where locomotives can use up to three times the amount consumed by yard switchers. These locomotives are expected to reduce nitrogen oxide and PM emissions by up to 80% while using between 20% and 40% less fuel.

The RP20BH locomotive uses two 667-hp diesel–generator sets and a bank of valve regulated lead acid batteries (700 VDC nominal, 600 Ah, and 500 kW).

Depending upon the power requirements (road or yard) the unit will either run on battery alone, one engine, or any combination of battery and engines. Having two smaller engines instead of a much larger single engine increases overall reliability and makes the locomotive easier to maintain, according to RailPower. The engines are skid-mounted for easy repair or replacement.

The specifications of locomotive Rp20BH is shown in **Table 5.4** [5.12].

Union Pacific Railroad began studies and tests of a RP20BH switcher locomotive in 2002 and has ordered 18 of these same locomotives.

References

5.1.Taketo Fujii, Nobutsugu Teraya, and Mitsuyuki Osawa "Development of an NE train." JR EAST Technical Review, # 4, p. 62-70

5.2. Takashi Kaneko, Motomi Shimada, Shinichiro Kujiraoka, Tetsuo Kojima" Easy maintenance and Environmentally-friendly Train Traction System" Hitachi Review Vol.53 (2004), No.1, p. 15-19

5.3. Yukio Arimori "Towards the future materialization of an environment friendly Railway" JR EAST Technical Review, # 4, p. 51-61

5.4. M. Osawa "NE train" Rail International, # 4, 2004, p.16-23.

5.5 The Associated Press. Japan to introduce the world's first hybrid train. International Herald Tribune. July 28, 2007

5.6. D.Lenhard et al. "Elektrische Ausrustung des triebzuges LIREX- Baureihe 618/619 fur DB Regio "Elektrische Bahnen." # 8, 2000, s.279-287.

5.7. C. Fischer "Triebzugfamilie CORADIA LIREX." Elektrische Bahnen, #8/9, 2002, s.289-296.

5.8 Technologies "Diesel-electric vehicles with energy storage" UIC 2003, Date created 10.09.2002.

5.9. "Electronic replace iron". IRJ, April 2004

5.10 "Electronic power supply system reduces operating cost." Innotrans 2004 Report.

5.11. B.Bradley Queen "The Green Goat hybrid locomotive" 2002.

5.12 RailPower product. http://www.railpower.com/products_gg.html

5.13. Bringshaw. "Hybrid shunter offers high fuel economy". IRJ, # 12, 2004

5.14 .F. Donnely, R. Cousineau, N.Horsley "Hybrid technology for the rail industry," p.1-5, 2004 ASME/IEEE Joint Rail Conference April 6-8, 2004, Baltimore, Mariland USA.

5.15 RailPower Technology Corporation. "Hybrid locomotives: better economics, better environment." ARB's Rail Symposium, July 13, 2006.7.15 5.15 Parry People Movers- Technology 16 January, 2006
http://www.parrypeoplemovers.com/technology.htm

5.16. Parry People Movers. PPM 80 Specifications
http://www.parrypeoplemovers.com/PPM80-spec.htm

6. HYBRID TRAMS, TRAINS AND LOCOMOTIVES WITH ENERGY STORAGE – LATEST DEVELOPMENTS.

6.1. Hybrid trams, trains and locomotives with energy storage and heat engines.

6.1.1. Hybrid tram with microturbine and flywheel energy storage (project)

A nine member European Consortium with European Commission funding has been developing an "Ultra Low Emission Vehicle - Transport using Advanced Propulsion" (ULEV-TAP) since 1996. The hybrid transport project is developing a series of configurations of a prime mover unit linked to an energy storage unit via a direct current link supplying the energy to the traction system as well as auxiliaries. The critical elements of the drive train encompass a high speed generator, driven directly by a gas turbine engine, a flywheel energy storage unit, power control and supervisory control electronics. The series of electric hybrid configurations allow for a much smaller prime mover unit for the vehicle and the ability to operate at peak performance. The benefits of lower emissions, higher system efficiency and bimodal operation are the main advantages of the ULEV-TAP system. A block diagram of the ULEV-TAP hybrid vehicle is shown in **Figure 6.1.**

ULEV-TAP is a hybrid with the Electro Mechanical Accumulator unit (EMAFER). The EMAFER is a high power flywheel energy storage device that is capable of storing and releasing considerable amounts of energy efficiently and instantaneously.

The design philosophy of EMAFER is to make full use of the high power capability of the flywheel. The high-energy efficiency and the long life cycle distinguish this system from the well-known electrochemical battery. The high speed system consists of a composite flywheel integrated with a synchronous permanent magnet motor and controlled by a bi-directional converter with high frequency power switches. The motor - generator has the power of 100 kW and weighs only 12 kg and has dimensions of 180 mm x 300 mm [6.1].

The new machine is fully air-cooled with an efficiency exceeding that of conventional technology. The machine may be operated as a generator when the high frequency output is rectified to DC and this energy is used for the feeding of auxiliary services. When the traction motors are used in braking mode as generators their energy is fed to the machine by means of inverter. The machine operates in this mode as a motor.

For ULEV-TAP this high speed generator is coupled directly to a gas turbine engine. Depending on the applications of the ULEV-TAP concept for future modern tram transit systems such as tram trains and tramway turbine-generators, characteristics can be optimized to provide a compact and efficient system integrated into the vehicle.

The gas turbine engine has a diameter of only 285 mm and the whole turbine-generator has an overall length of 660 mm with a total weight of 33 kg. The starting speed for the gas turbine is 29.000 rpm.

The components of the power system are build-up around a common DC-bus. Those components are: two DC-traction choppers based on IGBT technology, including line filters, chopper control and ventilation system. For further exploitation, AC-

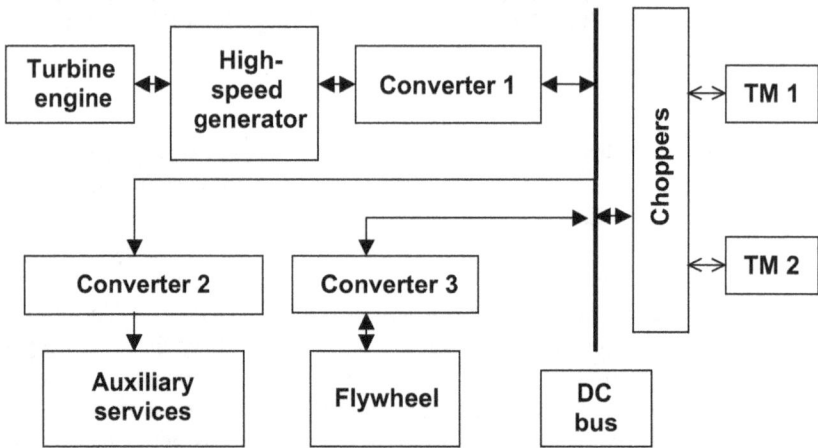

Figure 6.1 Block diagram of the ULEV-TAP hybrid tram

drives would be used. The converter 1 is used to adapt the rectified output-voltage of the high-speed generator to the DC-bus. This unit is based on IGBT technology. For gas turbine start-up, the directly-coupled generator is used as a motor. It is powered by an air-cooled IGBT start-inverter with a PWM switching frequency of 15 kHz. The converter 2 is used for the feeding of the auxiliary services.

6.1.2 Hybrid diesel train with ultracapacitor energy storage (Mitrac Energy Saver, Bombardier Germany).

Bombardier has modeled [6.2- 6.3] a 3-car articulated diesel-electric vehicle designed for a maximum speed of 120 km/h for local services with stops at 6 km intervals. The modeled vehicle has two 315 kW diesel power packs under the floor of the end cars, each with a power converter on the roof. The center car has a Mitrac Energy Saver with ultracapacitors on the roof which can store 5.4 kWh.

The diesel-electric train with energy storage has the following advantages:
• Better acceleration (the relatively light energy storage substantially increases the power).
• The decrease of the fuel consumption.
• The decrease of the emissions (especially CO_2).

The train has the ability to run several sections with a disconnected diesel engine (for example, in a tunnel or in a railway station building) or even move with a faulty diesel engine.

The fully charged energy storage (energy capacity is equal 5.4 kWh) allows the train to run about 2.8 km at a speed of 20 km/h. The comparison of the average accelerations to the speeds of 50 km/h and 100 km/h shows the advantages of the train with energy storage.

The return on the investment (ROI) for one 3-car diesel train with Mitrac Energy Saver can take from 1.4 to 4.5 years. This ROI is based on the following assumptions: 150000 km/year, 1.6 liter/km, 4.56 dollar for 1 fuel gallon, 8% interest rate.

6.1.3. British Green Goat switcher.

RailPower Technologies Corp., of Vancouver, Canada reached a supply and license agreement with Brush Traction Company under which the United Kingdom-based rail equipment supplier will exclusively manufacture Green Goat hybrid locomotives in the U.K. and Ireland.

The British Green Goat [6.4] is created using a Class 86 25kV electric locomotive. It is stripped to a bare shell with every component removed. About 60 electric locomotives Class 86 were built in 1965. This locomotive has a Bo-Bo wheel arrangement, power of about 3000kW, weighs 85t, nominal traction effort is 258 kN, maximum speed is 160 km/h and voltage is 25 kV, 50 Hz.

The braking system, traction motors, bogies, couplers, frame and other equipment are likely to be reused in the finished design. The transformer, internal cab equipment, pantograph and other parts are surplus and disposed of.

The modernized locomotive Class 86 is the first twin-cab Green Goat, with reconditioned controls and instrumentation, heater and air-conditioning fitted to the refurbished interiors. The "86" power controller is modified to provide step less adjustments of the traction effort while the direction selector, brake handle and instrument layout remain largely unchanged.

The rest of the bogy is adapted to make space for the battery cooling fans, a battery/engine pack (including generator), a compressor, traction motor blowers, and all necessary control cabinets and circuitry.

A fuel tank to supply the engine is fitted under the frame. The British Green Goat uses the Cummins QSK-19 engine. The engine and generator are in the centre of one side, with the engine positioned so that the side is equipped with filters to face the centre corridor. The engine is completely encased in acoustic insulation, together with a roof-mounted silencer to reduce noise.

Three electrical fans draw air over the roof-mounted cooler group. The brake frame is at the end of the engine for ease of maintenance.

The specifications of the locomotive British Green Goat are shown in **Table 6.1**.

Table 6.1 Main parameters of the hybrid locomotive British Green Goat

Traction horsepower (equivalent)	1200
Traction system	Battery and engine-generator
Wheel arrangement	Bo-Bo
Weight	72 tones
Locomotive control	Microprocessor, individual traction motor isolation
Maximum speed	100 km/h
Battery voltage	600VDC
Battery Type	Valve Regulated Lead Acid
Diesel engine type	Cummins QSK-19
Generator power	500 kW

6.1.4 Swedish Green Goat switcher

Swedish Train Technology (STT) will explore the market potential for RailPower's hybrid locomotive technology in Scandinavia [6.5] STT has converted a former Swedish State Railways T-43 diesel-electric built in the early 1960, the Nohab. The locomotive has Bo-Bo wheel arrangements, power of about 1000kW, a nominal traction effort of 210 kN, maximum speed of 100 km/h. The General Motors 567 series engine has been replaced by two banks of batteries, which are charged by a modern Cummins diesel. All the principal characteristics are the same as the RailPower Green Goat. The layout of the Swedish Green Goat includes several differences from the RailPower Green Goat. The engine-generator set of the RailPower's Green Goat is placed in front of the cab, while in the Swedish Green Goat it is placed in the rear of the cab.

6.1.5 Hybrid modernized switcher locomotives of Alstom Transport (France)

Alstom Transport has developed a hybrid switcher locomotive. [6.6-6.7]. The locomotive is the modernization of a former German diesel-hydraulic Bo – Bo locomotive class BR 212 (also known as V100).

Alstom has replaced the traditional diesel hydraulic propulsion with a battery propulsion system. The new propulsion system includes a nickel-cadmium (NiCd) battery, a 200 kW Deutz-engine diesel, a generator, and the traction inverter which feeds two three-phase asynchronous motors. Every motor has 213 kW of power. The locomotive can operate on a diesel-generator set or battery power alone, or with both sources together.

The generator charges the battery when needed, and the battery is kept permanently charged. The generator also can provide additional energy to the propulsion in peak times. The power of the locomotive in this mode is 550 kW. The electric motors work through a new split transmission, which is flange-mounted onto the cardan shafts used previously.

To balance the weight of the centre-cab unit, the diesel engine and auxiliary components are mounted on one side of the cab, and the batteries and traction inverter are mounted on the other.

The service mass of the locomotive is 64 ton; the top speed is 60 km/h. Starting traction effort is 195 kN, falling to 20 kN at a maximum speed.

The performance curve compares closely to a conventional diesel-hydraulic, with a maximum train weight of 3000 t.

Alstom estimates that the advantages of this hybrid locomotive are:
- The decrease of fuel consumption by 40%.
- The decrease of maintenance costs by 15%.
- The decrease in life –cycle cost by 30%.
- The decrease of noise at 15 dB(A).
- The decrease of emissions by 55%. Alstom planned to begin field testing this locomotive in April 2007.

The two-month long trials of the hybrid switcher started in Rotterdam on April 2009 and ended in June 2009.

The trials were successful and consistent with the Alstom forecast. Fuel savings were at least 40%, and the volume of carbon dioxide, nitrogen oxide, and particulate emissions was halved.

6.1.6 Hybrid turbine-electric switcher locomotive (Russia)

Russian hybrid locomotive is a modernized diesel switcher locomotive Class TEM 3. [6.8] The braking system, bogies, traction motors, couplers, frame and other equipment are kept from the original switcher while the diesel engine, main generator, and other parts are being removed.

The heart of the locomotive's power system is a natural gas fueled turbine.

The turbine turns an AC high speed generator (speed of rotation is 26000 revolution per minute), producing electricity. The switcher is equipped with the ultracapacitor's energy storage. Ultracapacitors are the energy storage means of choice in this application due to their higher efficiency, less maintenance, faster charge rate, wider temperature range, longer life cycle, and lower costs when compared with batteries. The locomotive has 6 traction motors, which are DC motors with separate excitation.

The main characteristics of the old diesel switcher included:
• wheel arrangement – 3o-3o
• engine power rating -880 kW (1200hp)
• service mass- 120 t
• maximal speed- 100 km/h
• continuous traction effort – 200 kN
• continuous rating speed -11km/h
• rated power of traction motor – 113 kW

The new features of the switcher include:
• gas turbine and high-speed generator
• DC traction motors with separate excitation
• ultracapacitor's energy storage,
• regenerative mode for locomotive with heat engine

6.1.7. GE hybrid diesel locomotive (USA)

Less than a decade ago General Electric (GE, USA) made public plans for the design of their first hybrid diesel-electric locomotive. [6.9-6.10]. GE planned that this project would last from January 2003 to December 2007.

This project is targeting a total of 20% of fuel usage reduction:
• 15% by capturing and storing of regenerative braking energy and additional
• 5% by optimizing the systems controls.

This project is very important for the USA and for the world's railroads. Today railroad modality in the USA is 2.5% of national fuel usage. The project's 20% fuel reduction gives a 0.5% impact on national fuel usage.

The energy storage should have a 1500 kW of continuous charge/discharge power and 1000 kWh of useable energy. Several comparisons of accumulator's batteries have been made – including two variants of the NiMH battery, Li-Ion battery, Na-NiCl$_2$ energy design and Na-NiCl$_2$ power design. As a result the Na-NiCl$_2$ battery has been selected.

The weight of the hybrid diesel locomotive is 207 ton. This weight is higher than the weight of the common GE Evolution Series locomotive ES 44AC (185 ton).

The benefits from the development of the new hybrid locomotive include:
• energy reduction (fuel savings of 750-950 million gallons a year).
• environmental impact (reduction of HC + NOx by 232 million lb a year).

On May 27th 2007 GE introduced it's first freight hybrid locomotive at the Union Station in Los Angeles, California. [6.11].

Before the GE hybrid locomotive is offered commercially, the engineering team will continue analysis on the new systems on-board the locomotive.

6.1.8 Hybrid High Speed Train (Great Britain)

On May 3rd 2007 Hitachi Europe unveiled "Hayabusa", which it claims to be Europe's first battery-assisted diesel-electric power car. This British High Speed Train (HST) power car is named after the Japanese word for a falcon. [6.12]

The parameters of the HST power car [6.13] are:
• power of rail-1320kW; wheel arrangement- Bo-Bo;
• mass-70,25 ton;
• maximum speed - 125 mph (200 km/h);
• fuel capacity - 4500 liters.
• cars use different engines, the Pachman Valenta 12RP200L, the Pachman 12VP 185, and the MTU 16V4000. The type of engine depends on the number of cars. Block diagram of HST is shown in **Figure 6.2**

The HST project is the next step in Hitachi's development of its hybrid technology. The prototype New Energy Train was described earlier in Chapter 4 of this book. This prototype has been tested in Japan since 2003 and JR-East put three hybrid DMUs based on the original prototype, into operation in July 2007.

The new power car uses a bank of 48 Lithium-Ion battery modules. Each module has energy capacity of 1kWh and weights 20 kg. With this new design, Hitachi is expecting a 20% cut in fuel consumption. Block diagram hybrid High Speed Train is shown in Figure 6.2.

To acquire equivalent power for current vehicles, the voltage of the DC composition (rectifier output) must be set to be around 1400 V. Since the voltage of battery banks was kept as 700V between the battery and the input of traction inverter is installed a stop-down/up chopper with IGBT transistors which is described in 2.2.3.[6.14]

Trials are being conducted of the prototype technology which was developed by Hitachi with its English partners Brush Traction as well as Network Rail and Porte-brook Leasing.

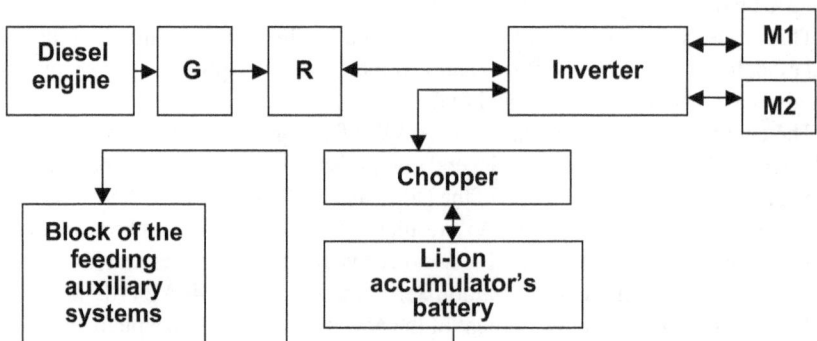

Figure 6.2. **Block diagram of England's hybrid High Speed Train**

The new hybrid HST car runs at speeds of up to 96 km/h. The car was tested as part of Network Rail's New Measurement Train (NMT). It was paired with a conventional HST power car which allowed a direct comparison of performance and energy efficiency. NMT required it to operate at speeds up to 200 km/h.

Test train which name is V-train 2 is undergoing running tests into the existing of the UK railway network.

Newly built power cars will have a better design and less powerful diesel engines and cooling groups. Hitachi is also expecting that the hybrid power car will have a reduced weight compared to existing cars, despite the battery modules.

The trials of Hayabusa power car equipped with onboard battery energy storage have been successful. The decision has been made that a new fleet of 200 km/h capable Super Express trains would replace existing IC 125 and IC225. The Super Express fleet will consist of electric, diesel and dual mode trains. The diesel and dual mode trains will use hybrid Japanese technology to reduce fuel consumption and emissions.

6.2 Hybrid trams, trains and locomotives with fuel cell energy storage.
6.2.1 Hybrid fuel cell tram (FULLTRAM project)

The long-term objective of the FULLTRAM Project [6.15] is to demonstrate the viability of a hybrid tram using a fuel cell and an on-board energy storage device. Producing the energy onboard can be achieved using fuel cell technology, with fuel storage on board the vehicle. To decrease the energy cost, on-board energy storage would be used to recover braking energy. The capital cost of such tram systems associated with the fuel cell engine, hydrogen storage and refueling station can be offset by the economy resulting from the absence of substations, overhead lines and associated high voltage equipment online and in the depot. Using overhead continuous electrification with the conventional trams requires grounding the current through the rails. Fuel cell trams don't need underground insulation and this is its big advantage. The principle of using a fuel cell and an on-board energy storage device is that this can be used for

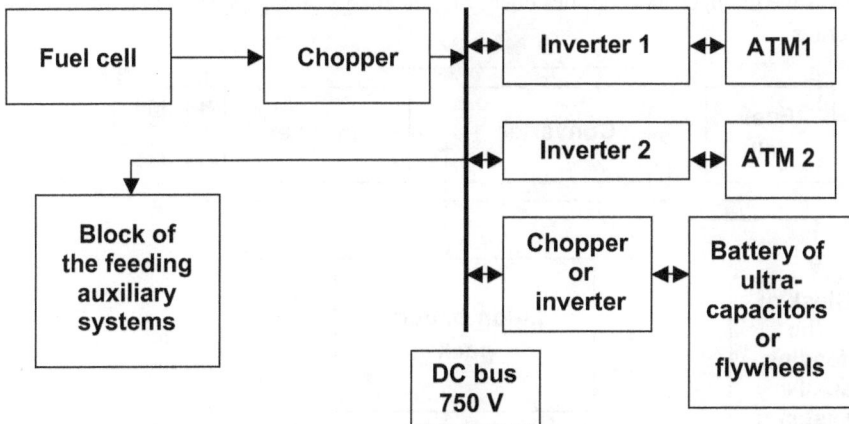

Figure 6.3 Block diagram of the hybrid fuel cell tram

the development of tram trains, avoiding the need for electrifying portions of existing lines. The removal of the overhead lines is also an advantage of the fuel cell tram.

The visual impact of the overhead lines becomes a concern for many cities, in particular those with a preserved historical centre. Several cities already expressed their interest for systems with partial or total removal of the overhead lines. Conventional tramways have great efficiency because they use the braking energy, which is transformed into electric energy by the motor which is used as a generator. This energy is used to supply the auxiliaries during the braking phase and also to power another vehicle, which can use it via the overhead contact line.

If we remove the overhead contact line, the braking energy is not reusable by another vehicle. To preserve the global efficiency of the tramway line, we have to store this energy in the vehicle and then use hybrid architecture. A bloc diagram of the hybrid fuel cell tram is shown in **Figure 6.3**.

The technology used for this onboard storage system can be flywheels, ultracapacitors or batteries. If the flywheel is used, an inverter is required to be placed between the energy storage and the power DC bus. If the battery or ultracapasitors are used, the inverter will be installed in the choppers.

6.2.2 World's First Fuel Cell Hybrid Railcar (East Japan Railway).

In Chapter 4.1 of this book we have provided a description of the world's first hybrid railcar (Japanese NE train).The Fuel Cell Hybrid Railcar is the modification of the NE train [6.16].

The block diagram of the Diesel-electric Hybrid Railcar is shown in **Figure 5.1**;

The block diagram of the Fuel Cell Hybrid Railcar is shown in **Figure 6.4**.

The diagrams (**Figure 5.1** and **Figure 6.3**) have a difference. The engine and generator of the NE train (**Figure 5.1**) are replaced with fuel cells (**Figure 6.4**).

Solid polymer type fuel cells are implemented for energy storage. The converter in **Figure 5.1** is used as the rectifier in the traction regime and as the inverter in the regenerative regime, when it is used for engine braking. The converter in **Figure 6.4** is used as the chopper that supplies energy to the energy storage system and the traction inverters.

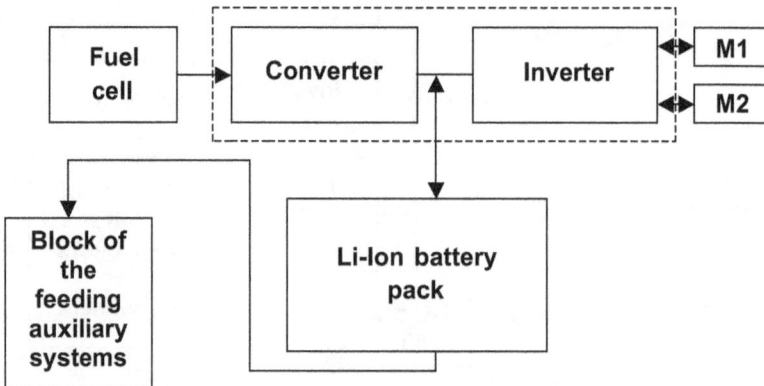

Figure 6.4 Block diagram of the hybrid fuel cell train (Japan)

The capacity of the fuel cell is 2*65 kW. The hydrogen tank capacity and pressure are approximately 270 liters and 35 MPa. It uses a Lithium-ion type 19 kWh battery. The control system directs electric power from both the fuel cell and the storage battery when accelerating, and saves the electric power produced by the regenerative brakes in the storage battery when braking.

Looking at the image of the fuel cell railcar one can observe that the hydrogen tank and the fuel cells modules are placed under the car body and the Li-ion batteries are placed on the roof.

The new drive train has the following advantages:
• energy consumption reduction (fossil fuel economy)
• the emissions. reduction

Original trials of the first fuel cell hybrid railcar began in July of 2006. The tests were to confirm the new drive advantages, the fuel cell performance, and the features of hydrogen supply system.

The world's first fuel cell powered hybrid train was tested in 2006. In the spring of 2007 this hybrid train started test runs on operational lines. The latest tests confirmed that the hybrid train could reach the speed of 100 km/h.

6.2.3 Fuel cell train project of Japan Railway Technical Research Institute (RTRI).

The project started in Japan in 2001 [6.17] under the name "Basic study on application of fuel cells to the traction systems of railway vehicles." The project was planned to run for the next 6 years. The propulsion conception is a hybrid (fuel cell–battery) concept. The fuel cell that was chosen is based on PEM FC (proton exchange membrane fuel cell) technology.

The main diagram of the RTRI fuel cell hybrid train is shown in **Figure.6.5.**

The features of the power circuits include:

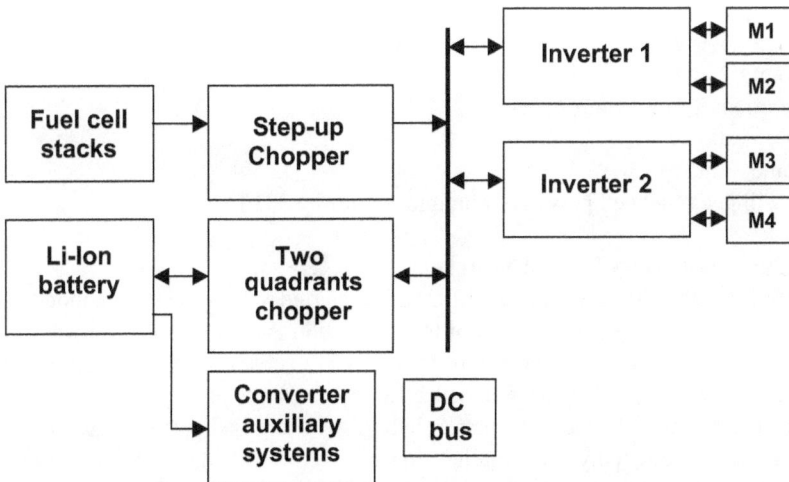

Figure 6.5 Block diagram of the RTRI hybrid fuel cell train (Japan)

• choppers are implemented between the fuel cell stacks and accumulator's battery
• Choppers are built using GTO thyristor technology
• Both traction motors are fed from the same inverter.

The important task for this research was selecting a fuel base for the future fuel cells. There were 3 candidates considered (pure hydrogen, liquid natural gas (LNG) and methanol). LNG and methanol could be implemented but would need utilizing special reformers to chemically generate the hydrogen.

The first car of the set has two power bogies (4 traction motors), converters and an energy source. The second car has fuel cell stacks, a reformer and a fuel tank. The estimated value of the total power of the set is 800 kW (fuel cells should generated 600 kW, the accumulators or ultracapacitors should generated the additional 200 kW). The estimated weight of two cars is 70 ton. The average range between fuel refilling should be 400 km, top speed – 120 km/h. The estimated fuel consumption is about 100 kg per day.

The initial research was focused on the development of a scale prototype unit. This unit used pure hydrogen and air for power generation and an electric train bogie (equivalent to that used for an actual electric commuter train). The bogie was set up on a vehicle test bench. In 2001 the bogie was tested with PEMFC that had the power of 1 kW and it reached the speed of 30 km/h. In 2003 the bogie was tested with PEMFC that had the power of 30 kW and it reached the speed 50 km/h. The module with 8 fuel cell stacks and the total power of 150 kW (net output power is 120 kW) was manufactured by US-based Nuvera Fuel cells. The module was mounted onto a vehicle and tests were conducted on the test track at RTRI. On October 18th 2006 RTRI announced that the institute had successfully conducted a test run of the railway vehicle powered by a fuel cell.

The test vehicle was equipped with the fuel cell power module, a high-pressure hydrogen tank system, the asynchronous traction drive and was powered and driven by a fuel cell system in the test run.

The characteristics of the Forza Rail Power Module (RPM) (Nuvera Company) are as follows:

• Voltage - 600 V DC,
• Current - 250 A,
• Dimentions:1,6m * 1,2m * 1.5m
• Masse -1500 kg
• Volume -3 m^3.

RTRI will get a fuel cell powered train into service by 2010.

6.2.4 Fuel cell-battery Hybrid Switcher

On July 14th 2006, Vehicle Projects LLC (USA) started to develop the the modernized hybrid switcher [6.18]. The object of modernization is a famous Rail Power hybrid switcher (see section 5.3). The aim of this work is to change the diesel-generator set of the rail power's switcher to a fuel cell system.

The main diagram of the fuel cell –battery hybrid switcher - is shown in **Figure 6.6**.

The fuel cell - battery hybrid switcher doesn't need fossil fuel and there are no emissions. It has more advantages in comparison to the common switcher than the hybrid switcher with a big battery and a small diesel-generator set.

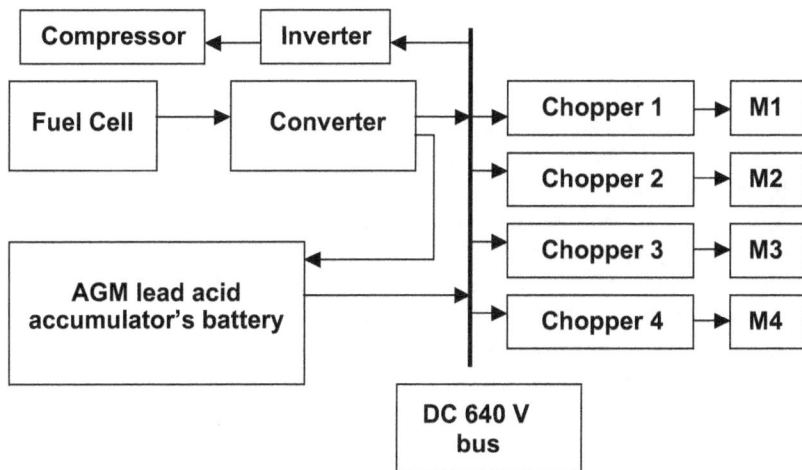

Figure 6.6 Block diagram of the fuel cell hybrid switcher (USA)

At first Vehicle Projects LLC has proposed to use two 150 kW proton exchange membrane fuel cell power plants. This power plant is developed by Vehicle Project LLC and uses Nuvera Fuel Cell's FORZA power modules.

Now Vehicle Projects LLC decided to use the fuel cell power plants which are based on Ballard Power Systems' Mark 902 P5, 150 kW proton exchange membrane (PEM) power modules. These modules were used for 27 Citaro buses to the Clean Urban Transit in Europe (CUTE) program and for 9 Citaro buses for Iceland, Australia and China. These modules have over 1.5 million km of proven reliability over 3 years. The converter in the output of fuel cell source is a DC-DC step-up chopper. The base parts of the Rail Power Switcher Drive system have not been changed.

Vehicle Projects LLC decided also to use Dynetek's 350 bar composite cylinders also used on the Citaro bus program. Citaro buses were equipped with gaseous hydrogen cylindres and carry from 40 to 50 kg of hydrogen fuel in eight to 11 cylinders at a pressure of 350 bar.

An operational hydrogen fuel cell switch locomotive was unveiled at BSNF's Topeka System Maintenance Terminal in June 2009. After this demonstration the switcher went to the Transportation Test Center in Pueblo, Colorado. It is planned to go into service in the Los Angeles Basin.

References
6.1 ULEV-TAP newsletter # 2 of August 2000.
http://www.ulev-tap.org/ulev1/opening.html
6.2.M.Steiner, S.Pagiela "Energyspeicher in Schinenfahrzeugen". Eisenbahn Technische Rundschau (ETR # 4, 2005, s.207-214)
6.3. H.Steineregger "Einsatzmoglichkeiten von Energyspeichern in dieselbetriebener Schinenfahrzeugen". Eisenbahningenieur,(%$), # 12,2003, s. 17-19.
6.4.Chris Gilson " Battery are included", Rail 529, December 21 2005-January 3 2006, p.31-33.

6.5.Swensk tagteknik (Swedish Train Technology). Hybridteknik I nya tidens lok (Hybrids bringing a new era for locomotives).
www.stt-train.se/pdf/hybrid_www.pdf
6.6 Hybrid shunter conversion. Railway Gazette International 01 November 2006
6.7 InnoTrans 2006: The big diesel-locomotive rendezvous. Railway Technical Review 4 2006 page 32
6.8 ГАЗЕТА "ГУДОК" 25.04.2005 А.Логинов
The Russian newspaper "Gudok", 25.04.2005, A. Loginov.
6.9 Robert D. King, Lembit Salasoo "21st Century Locomotive Technology (locomotive system tasks). GE Global Research. April 2006.
6.10. GE imagination at work. Hybrid locomotive (a product of Ecomagination). General Electric Company 2005.
6.11 Green Car Congress 23 May 2007 GE to introduce Hybrid Road locomotive
6.12. "Hybrid High Speed Train unveiled". Railway Gazette International, 04 May 2007.
6.13 British rail class 43 (HST) rail http://en.wikipedia.org/wiki/British_Rail_ Class_43_%28HST%29
6.14 Kazuo Tokuyama, Motomi Shimada, Keyashi Terasawa, Kiyashi Kaneko " Practical Application of a Hybrid drive system for Reducing Environmental Load" Hitachi Review, vol.57 (2008), #1, p. 23-27
6.15. T.Montanie, J.P. Moskowitz "Hybrid fuell cell tramway (Fulltram project) http://www.waterstof.org/20030725EHECO4-155pdf
6.16 JR-East: press release> Development of the Word's first fuel hybrid railcar www.jreast.co.jp/e/press/20060401/index.html - 30k
6.17. RTRI Fuel Cell Train. -FC Application to the Traction System. Specification of the FC train. JR. Railway Technical Research Institute
www.hydrail.com/docs/kondo.pdf
6.18 Arnold Miller "Fuel cell hybrid switcher locomotive." Project status as of 4 Dec 2006 www.fuelcellpropulsion.org

7. THE ANALYSIS OF THE MODERN HYBRID TRAINS AND HYBRID SWITCHERS.

7.1 The features of the hybrid diesel multiple unit.

The main characteristics of the hybrid diesel electric multiple unit (HDEMU) is listed in **Table 7.1.**

A summary of the above characteristics of the real HDEMU include:

• Power rating of diesel engines is 330-338 kW.

• Power rating of the main generators is 180-320 kW.

• Energy of the energy storage is 6-10 kWh.

• Cars weight is 23-26 t

All three HDEMUs use the asynchronous traction drive, which became a standard approach. All new electric and diesel lightweight cars from Europe and Japan have asynchronous traction motors. The asynchronous traction drive provides the ability to go from traction conditions to regeneration very easily. Asynchronous traction drive has high efficiency, small dimensions and is lightweight. All HDEMUs have implemented converters with IGBT.

The auxiliary converters for each of the three HDEMUs are connected to the same DC link as the power converters. This is an optimal configuration as it creates the ability to use the braking energy for the auxiliaries.

LIREX has the choppers and a brake resistor, the NE car does not have the resistor braking. When running at a constant speed, the brake system is regenerating the power for the NE car and this power is absorbed when using engine braking.

Table 7.1 The main characteristics of the hybrid diesel electric multiple unit (HDEMU)

Name of HDEMU	NE train, diesel railcar E991 (Japan)	Diesel Train LIREX	Diesel train with Mitrac Energy Saver (project Bombardier)
Quantity of cars	1	6	4
Quantity of diesel engines,	1	4	2
Power of diesel engine, kW	330	338	315
Power of main generator	180	322	
Weight of car., t	26	23	3 cars train weight 100t
Quantity of traction motors	2	4	
Power of traction motors, kW	95	190	
Type energy storage	Lithium Ion Battery	Flywheel	Ultracapacitor (Mitrac Energy Saver)
Energy of energy storage, kWh	10	6	5.4
Power of energy storage, kW	250	350	

The control system of the HDEMU should have in its structure the energy management block that controls the energy supply/distribution of both the diesel engine and the energy storage unit.

7.2. New features of hybrid switcher Green Goat 20, Green Kid 10B

The GG20 and GK10B hybrid locomotives are quite different from other switchers developed in the earlier days. The motors of these locomotives can be fed from a diesel engine and a battery bank at the same time. Diesel engine is rarely used as a single source. This switcher works most of the time like the battery-powered electric locomotive, only this locomotive uses diesel-generator to charge the battery bank. As with other battery-powered electric locomotives, the GG20B and GK10B also have electric drive elements, including battery banks (lead-acid), choppers and direct current motors. In contrast to other battery-powered electric locomotives these switchers have an autonomous diesel-generator that is charging the battery bank. This way the capacity of the battery bank is sustained at 80% from a nominal level. Such approach increases the reliability and service life of the batteries. The early models of the battery-powered electric locomotives did not have additional source of charge on board and had a limited range.

The basic principles used in design of these hybrid switchers are similar to that of the hybrid autos. Both hybrid locomotives and autos use a combination of a heat-engine and a rechargeable energy source (generally electric batteries). An autonomous diesel-gener-

Table 7.2 Comparison of power ratings for both heat-engine and electric motors of hybrid switchers, trains, buses, and autos

Hybrid Type	Hybrid Model	Power rating of heat-engine (PRHE) in kW	Power rating of electric motor (PREM) in kW	DOH PREM / PRHE+ PREM
Locomotive	GK10B	90	360	0.8
Locomotive	GG20B	200	720	0.78
Locomotive	RP 20BH	1000	500	0.33
Locomotive	Hybrid switcher of Alstom	200	426	0.68
Train	New Energy Train (NE, Japan)	330	250	0.43
Bus	Orion 7	197	36	0.15
Auto	Honda Civic Hybrid	62.5	9.85	0.14
Auto	Toyota Prius	55.9	49.2	0.47
Auto	Ford Escape Hybrid	97.7	70.0	0.5

ator is the charging source that is used in both hybrid locomotives and autos. An electric motor helps a heat-engine in the hybrid locomotives, as it does in the hybrid autos.

Many diesel switchers idle about (82-86) % of their operational time (in correspondence with AAR). [7.2] Therefore they work only (14-18) % of their operational time. GG20B and GK10B hybrid locomotives didn't idle. They could be charged during the time when they are out of operation. Long charging time allows using lower charging current. Therefore the power of diesel-generator used to charge the equipment can also be less.

Hybrid autos idle for only 22-25% of their operational cycle (in correspondence with city cycle SAEJ2711). Therefore the power of ICE has to be greater than the power of the electric motor. In hybrid systems, it is useful to define a variable called the "Degree of Hybridization (DOH)" as follows: DOH = electric motor power/ electric motor power + heat engine power.

Table 7.2 lists power ratings of both heat engines and electric motors of hybrid switcher locomotives, hybrid train, hybrid bus, and several hybrid autos.

As it appears from **Table 7.2**, DOH of the hybrid locomotives Green Goat and Green Kid is more than that of the other modern hybrid vehicles. The hybrid switcher locomotives GK10B and GG20B are more closely related to the battery powered locomotives. At the same time Road Switcher RP20BH is different from these two switchers as it is more closely related to the other modern hybrid vehicles.

Remanufactured hybrid switchers have the following advantages compared to the original diesel-powered switchers:
• Emission reduction
• Noise reduction
• Constant operation readiness

Remanufactured hybrid switchers have the following advantages compared to the earlier models of battery powered locomotives:
• Utilize a maintenance-free battery bank with a long life of service
• Utilize a diesel-generator as a source of charge for the battery bank
• Generally greater range.

There is an environmental disadvantage in using GK10B and GG20B locomotives compared to the battery powered locomotives as they have a small diesel engine on board. The new GG20B locomotive with fuel cell doesn't have the same disadvantage.

A microprocessor's control system provides the energy for the traction motors. This system is based on the estimation of the adhesion limit.

Therefore the starting traction effort of these locomotives increases by 35% in comparison with the starting tra5ction effort of the switcher diesel locomotive before remanufacturing.

7.3. A new perspective for hybrid switchers

The new technology for remanufacturing of the switchers has been a big success in protecting the environment and promoting railroads. The battery banks deployed on the switchers made it possible to operate in emission-free mode. The lead acid battery used has a power-to-weight ratio of 31 W/Kg, which is similar to the ratio of the traditional battery. The weight of the battery is about 20% of the locomotive weight;

however, using the maintenance free technology opens a whole new perspective. This technology allowed the remanufacturing of many switchers with DC drive which are still in existence today.

New locomotives will be designed and developed tomorrow:
• Micro turbine power generator or fuel cells will replace the diesel-generator used today.
• DC drive will be replaced by an AC asynchronous drive.
• Lead battery will be replaced by a better battery.

Implementing the asynchronous traction drive for the next generation switchers provides the following benefits:
• Regenerative braking allows the ability of recuperating the energy and making the locomotive completely noiseless.
• Decreasing energy loss by the continuous transitions from acceleration to braking regimes.
• Decreased expenditures for maintenance and repair
• Increasing the reliability of the locomotive.

The on board micro turbine generator recharges the battery packs on the switcher while it is out operation. The common micro turbine is smaller, lighter, quieter, and cleaner than traditional diesel engines. It has no gearbox, no pumps, and no mechanical subsystems. There is no need for oil, lubricants or coolants. The micro turbine consumes less fuel than traditional diesel engines. Even today there are micro turbines available with power ratings from 30 kW to 250 kW.

Fuel cells are used to charge the main battery instead of diesel-generator which provides a conversion of the hybrid switcher to ZEV (zero emission vehicles) switcher. As an example, the fuel cell for the locomotive can be used the fuel cell-battery road switcher. (See section 6.2.4)

The advantages of the micro turbine over the fuel cell for switcher locomotive are following:
• The micro turbine can use the present system of fuel distribution.
• The life time of the Capstone micro turbine is 20,000 hours; the life time of the PEM fuel cell is less.

7.4 The features of the regenerative braking on hybrid switchers.

The switchers go through many "start-stop" cycles during daily operation. The number of these shifting cycles can be equal to 100. If n =100, equation (14) can be rewritten as:

$$K_{doc} = 1 - q * \eta_{ES} / q + 1 \tag{23}$$

The equation for the coefficient of the return of energy k_{roe} can be written as:

$$k_{roe} = q * \eta_{ES} / q + 1 \tag{24}$$

It is possible to show that the value q in the case of the switcher on average is equal to:

q = 12,5 $*$ v^2/S, where: (25)

v – speed of locomotive [m/s], S- the way of start-braking cycle [m]. The calculated result of q by v = 2, 3, 4, 5 [m/s] as listed in **Table 7.3**

Table 7.3 Value q variation from speed and braking distance of hybrid switcher

Speed, m/s	Braking distance		
	250 m	500 m	750 m
2	0.2	0.1	0.08
3	0.45	0.225	0.15
4	0.8	0.4	0.27
5	1.25	0.625	0.416

Suppose that the effective regenerate braking is the braking with $k_{roe} > 10\%$. Formula (24) lets us calculate the value q that correspondences to this condition.
If $\eta_{ES} = 0.5$, q should be more than 0.25.
If $\eta_{ES} = 0.7$, q should be more than 0.16.
Table 7.3b shows the data on speed and braking distance when $\eta_{ES} = 0,7$ and q > 0.16 (the value q is showed face bold).

Table 7.3 b

Speed, m/s	Braking distance		
	250 m	500 m	750 m
2	**0.2**	0.1	0.08
3	**0.45**	**0.225**	0.15
4	**0.8**	**0.4**	**0.27**
5	**1.25**	**0.625**	**0.416**

Table 7.3c shows the data on speed and braking distance when η_{ES} =0.5 and q > 0.25 (the value q is showed face bold).

Table 7.3 c

Speed, m/s	Braking distance		
	250 m	500 m	750 m
2	0.2	0.1	0.08
3	**0.45**	0.225	0.15
4	**0.8**	**0.4**	0.27
5	**1.25**	**0.625**	**0.416**

This calculation shows that for speeds greater than 4 m/s k_{roe} is considerably more than 10%.

The authors of article [7.4] confirm that for switchers with DC drives the regenerative braking does not provide great benefits. This will be true for a diesel-electric locomotive like GP-10 (which should be modernized). In case of the new rail vehicles with asynchronous traction drive we are sure that the hybrid rail vehicles should utilize the regenerative braking system and this is the right technology to employ to gain great benefit. Current implementations of regenerative braking in several real hybrid trams and diesel trains confirm this conclusion.

7.5 Modern energy storage systems

Energy storage became a critical element of the modern hybrid vehicle. Today we have a variety of energy storage systems on the market. The battery banks, flywheels, and ultracapacitors are the best devices for rail hybrid vehicles.

Battery banks are used as energy storage for diesel and fuel cell variants of the NE train (Japan) (Li-Ion battery), for modernized hybrid switch locomotives (USA) (AGM VRLA battery), for the fuel cell-battery hybrid switcher (USA) (AGM VRLA battery), and for the hybrid GE diesel locomotive (USA), (Na-NiCl$_2$ battery).

Flywheels are used for hybrid DEMU Lirex (Germany), hybrid trams PPM 50 and PPM 80 (England}, and the hybrid tram with microturbine and flywheel energy storage (Germany and European Consortium).

Ultracapacitors are used for LVR with ultracapacitor energy storage (Mannheim, Germany) and for hybrid turbine-electric switcher locomotive (Russia).

The characteristics of the modern energy storage systems are shown in **Table 7.4**:

Specific energy is calculated as: $S_E = A_{ES} / m_E$ (26)

Specific power is calculated as: $S_p = P_{ES} / m_p$,   (27)

where: m_E – mass with minimum energy requisite (kg),
         m_p – mass with minimum power requisite (kg),
         S_E – specific energy (kWh/kg),
         S_p – specific power (kW/kg)

Table 7.4 contains a comparison list of parameters for the real energy storage systems using the above formulas, the numbers in bold were obtained from the past publications describing non-conventional energy storage systems.

The diagrams of specific energy and specific power distribution in different kinds of energy storage systems are shown in **Figure 7.1** and **7.2**.

Looking at the diagrams, it's easy to notice that the Li-Ion battery has the best specific energy and the ultracapacitor has the best specific power. The devices that have higher specific energy have lower specific power and vice versa.

Energy and power that could be stored by different energy storage systems are shown in **Table 7.5**. EORB and PORB on freight diesel locomotives and freight trains are calculated by formulas (21) and (22). EORB and PORB on four other vehicle types are calculated by formulas (19) and (20).

Table 7.4 Comparison table of modern energy storage systems

Energy storage family	Vehicle with ES	Country	Energy of ES (kWh)	Mass of ES (kg)	Specific energy (Wh/kg)	Specific power (kW/kg)
AGM VRLA battery	Green Goat switcher	Canada	768	22500	34.1	0.2
Li-Ion battery	Tram with accumulator's battery	Japan	33.0	1160	28.4 **120**	0.8 **1.2**
NiMH battery	Car Toyota Prius	Japan	1.78	53.3	33.55 **70**	0.27 **0.42**
ZEBRA battery (Sodium Nickel Chloride) Swiss	Bus	USA	96.0	1188	80.8 **100**	0.15 **0.18**
Flywheel	DEMU Lirex	Germany	12.0	2600	4.6 **14.0**	0.35 **1.5**
Ultracapacitors Mitrac Energy Saver	Tram Mannheim	Canada, Germany	0.85	477	1.78 **6.0**	0.62 **6.0**

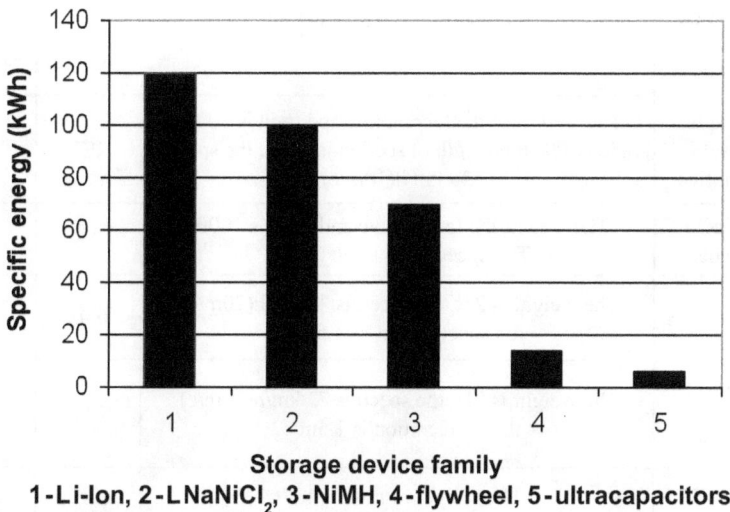

Figure 7.1 Specific energy of different energy storage systems

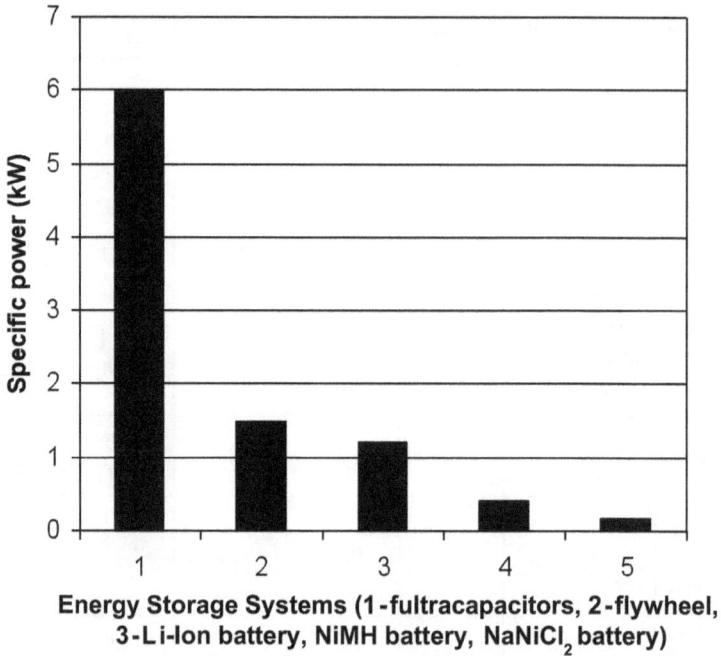

**Energy Storage Systems (1-fultracapacitors, 2-flywheel,
3-Li-Ion battery, NiMH battery, NaNiCl$_2$ battery)**

Figure 7.2. Specific power of different energy storage systems

Table 7.5 Brake energy and brake power for different rail vehicle types

Rail vehicle	The conditions of calculation	Energy (kWh) of energy storage	Power (kW) of energy storage
1. Freight diesel locomotive	The weight of the locomotive and train is 3000t, grade is 1%, the length of section is 5 km, the speed is 36 km/h(10m/c)	171	206.3
2. Hybrid switcher	The weight the locomotive and train is 1000t, The speed is 36 km/h (3m/c)	0.8	960
3. Tram-train	The weight is 25t, the speed is 72 km/h(20m/c), the acceleration is 1.5m/c2	0.98	264
4. Railcar (USA) Project	The weight is 91t, the speed is 72 km/h(20m/c), the acceleration is 1.5m/c2	2.93	792
5. DEMU Lirex	The weight of train is 137t, the speed is 160 km/h (45m/c)	25.62	2196

The energy and power of energy storage systems are calculated by formula:

$$A_{ES} = E_{RB} * \eta_{ES}$$

$$P_{ES} = P_B * \eta_{ES}$$

The most brake energy that could be stored appears to be on a freight diesel locomotive with the weight of 3000 ton. The greatest brake power comes from the DEMU Lirex at a speed of 160 km/h. The hybrid switcher has the least brake energy due to the low speed of the switcher which is about 2-3 m/c.

The USA railcars [7.3] have the strictest requirements for "crashworthiness", and therefore have much greater weight compared to their European counterparts. This railcar's weight is significantly higher than the weights of the other railcars. The tram-train and the Lirex cars (6 cars weigh 137t) are lightweight vehicles. These cars don't satisfy the requirement of the criterion "crashworthiness" defined by the US Federal Railroad Administration. This criterion requires that "all passenger equipment shall resist a minimum static end load of 800,000 pounds(362,870 kg) applied on the line of draft without permanent deformation of the body structure."

We will calculate the actual mass of each individual energy storage system described in **Table 7.5.**

$$m_E = A_{ES}/ SE ; \ m_p = P_{ES}/ S_p$$

We assume that the outputs of all energy storage systems are connected to the DC link through IGBT converters.

The results of calculations are shown in **Table 7.6.**

As for the battery pack, we have calculated that the energy value of the battery is 5 times greater due to the assumption that the battery is discharged at 20% when the regenerative braking starts. The battery SOC (State of Charge) is continually changing and is kept above 20%.

The lowest mass of energy storage system is shown in bold letters. During this calculation the specific mass of the converter was equal to 2 kW / kg.

Only in the case of the freight locomotive and the train, the mass which satisfies energy requirements is more than the mass which satisfies power requirements. It other cases it is vise versa.

The next calculation confirms this last conclusion.

Let's compare the masses with minimum energy and power requirements.

$$m_E = A_{ES}/ S_E$$

$$mP = A_{ES} / S_p * 3600/t , \text{ where:} \tag{28}$$

m_E - mass which corresponds to energy requisite (kg),
m_p - mass which corresponds to power requirements (kg),
S_E - specific energy (kWh /kg),

Table 7.6 The actual mass of energy storage systems for five different types of hybrid rail vehicles

a. Hybrid freight locomotive with the train

The ES family	Mass with minimum energy requested	Mass with minimum power requested	Mass of power converter	Total mass of ES
Li-Ion battery	7125	1030	618	**7743**
NiMH battery	12214	2992	618	12832
NaNiCl$_2$ battery	8550	6864	618	9168
Flywheel	12214	686	618	12832
Ultracapacitors	28500	206	618	29118

b. Hybrid switcher

The ES family	Mass with minimum energy requested	Mass with minimum power requested	Mass of power converter	Total mass of ES
Li-Ion battery	33	4000	480	4480
NiMH battery	57	11425	480	11905
NaNiCl$_2$ battery	40	26650	480	30130
Flywheel	57	640	480	1120
Ultracapacitors	166	160	480	**646**

c. Tram-train

The ES family	Mass with minimum energy requested	Mass with minimum power requested	Mass of power converter	Total mass of ES
Li-Ion battery	41	220	132	352
NiMH battery	70	629	132	761
NaNiCl$_2$ battery	50	666	132	798
Flywheel	70	176	132	308
Ultracapacitors	163	44	132	**295**

d. Project of railcar (USA)

The ES family	Mass with minimum energy requested	Mass with minimum power requested	Mass of power converter	Total mass of ES
Li-Ion battery	122	3300	396	3696
NiMH battery	210	1886	396	2282
NaNiCl$_2$ battery	145	4400	396	4796
Flywheel	210	528	396	924
Ultracapacitors	488	132	396	**884**

e. HDEMU Lirex (Germany)

The ES family	Mass with minimum energy requested	Mass with minimum power requested	Mass of power converter	Total mass of ES
Li-Ion battery	1068	9150	1068	10248
NiMH battery	1830	26140	1068	27238
NaNiCl$_2$ battery	1295	73200	1068	74298
Flywheel	1830	1464	1068	**4392**
Ultracapacitors	4270	366	1068	5734

S_P - specific power (kW/kg), t - time (s).
Let's find the condition when $m_P > m_E$

$$A_{ES} / S_P * 3600/t > A_{ES} / S_E \quad S_E / S_P * 3600/t > 1 \text{ and}$$

$$t < SE / SP * 3600 \tag{29}$$

This calcultion shows us that
Li-Ion battery : $t < 360$ s,
NiMH battery : $t < 600$ s
NaNiCl$_2$ battery: $t < 200$ s
Flywheel: $t < 33.6$ s
Ultracapacitors: $t < 3.6$ s.
The braking time of 200-600 s is not possible to achieve on the rail vehicles.
   In case of rail vehicles stop braking, the minimum mass is the mass which satisfies power requirements.

Table 7.7 The mass of energy storage if we use Li-Ion accumulator's battery and ultracapacitors together

Rail vehicle	Mass with minimum energy requested (accumulator's battery)	Mass with minimum power requested (ultracapacitors)	Total mass of ES (one kind of ES devices)	Total mass of ES (Li-Ion accumulator battery plus ultracapacitors)
Freight locomotive	7125	206	**7743**	7949
Hybrid switcher	33	160	**646**	673
Tram-train	41	44	295	**217**
Railcar (USA)	122	132	884	**650**
DEMU Lirex	1068	366	4392	**2532**

Table 7.7 in correspondence with **Table 7.5** and **Table 7.6** shows us the total mass of energy storage systems if we use only one kind of ES devises (Li-Ion battery packs, ultracapacitors or flywheels) or we use Li-Ion battery packs and ultracapacitors together. In both cases we included the mass of the converter. The minimum of the total mass of the storage system is shown in bold.

From **Table 7.7** follows that if we use Li-Ion battery and ultracapacitors together:
• the total mass of the storage system is equal 217 kg for tram-train, 650 kg for railcar (USA), and 2532 kg for DEMU Lirex. These masses are the minimum masses for these situations.
• in the case of the speed stabilization on a slope or grade, for the hybrid diesel locomotive and the freight train, the best option is to choose battery packs as energy storage devices.
• In the case of the hybrid switcher locomotive, the best option is
to choose ultracapacitor as the energy storage device.
• In the case of stop braking of hybrid rail vehicles (except the hybrid switchers) the best option is to combine use of two energy devices, battery packs and ultracapacitors together.

References
7.1 Yukio Arimori "Towards the Future Materialization of an Environment friendly Railway" JR EAST Technical Review # 4, 2004, p. 51-61
7.2 F. Donnely, R. Cousineau, N.Horsley "Hybrid technology for the rail industry", p.1-5, 2004 ASME/IEEE Joint Rail Conference April 6-8, 2004, Baltimore, Mariland USA.
7.3 Paul E.Kaufmann, Jurgen Schulte "Feasibility study of a series hybrid diesel multiple unit Railcar, p.1-9. http://www.kko.com/sprc/research/hybridsprc.pdf7
7.4. Miller A.R., Peters, J. "Fuelcell hybrid locomotives: application and benefits" Rail Conference 2006, Proceedings of the 2006 IEEE/ASME Joint. Publication date: 4-6 April 2006 Pages: 287-293 ISBN: 0-7918-4203-7

8. CONCLUSIONS

The above analysis contributes to the following conclusions:

• Modern rolling stock includes the conventional rail vehicles, dual-mode rail vehicles, DC traction units with the energy storage, and hybrid rail vehicles. The hybrid rail vehicles should be classified in compliance with the UN definition.

• Modern energy storage, high power converters, main generators, and traction motors promote the creation of the hybrid rail vehicles with high power efficiency.

• Regenerative braking is an essential technique to be used to boost the energy efficiency of hybrid rail vehicles. The regenerative braking on hybrid rail vehicles with heat engines is theoretically and practically proved.

• The efficiency of the components of the traction converters, main generators and traction motors has considerable influence on the value of energy which is saved by hybrid rail vehicles. That is why hybrid rail vehicles should use the components with high efficiency: generators and motors with permanent magnets, the converters with Insulated Gate Bipolar Transistor (IGBT), battery packs and ultracapacitors with less internal resistance.

• The battery packs, flywheels, and ultracapacitors are successfully deployed today as energy storage devices in the different kinds of rolling stock. Today we don't have a single universal solution for different kinds of rolling stock.

• In the case of the speed stabilization on a slope or grade, the hybrid diesel locomotive and the freight train, should use battery packs as rechargeable energy source. The hybrid switcher locomotive should deploy ultracapacitors as the rechargeable energy source. The hybrid switcher has a low speed, low braking energy and very short braking time. In the case of stop braking on hybrid rail vehicles (except the hybrid switchers) two combined rechargeable energy sources should be used: battery packs and ultracapacitors together.

• The methods for reducing electric energy loss that were used earlier for conventional rolling stock, such as the reduction of the weight, the reduction of the rolling resistance, implementing of the optimal control systems, could be also used with success on the hybrid rail vehicles.

• Energy storage system deployment on urban DC traction units will increase power efficiency, reliability and safety.

• Same converters, traction motors and auxiliary drives are used today for electric rolling stock and for rolling stock with heat engines. Multi system and dual-mode locomotives and trains exist today. Next step will be a development of multi system and dual-mode hybrid trains. Alstom Cardia Lirex train is an example of modern two system and dual-mode hybrid train.

• The considerable small mass and size of the power source with a heat engine allows for building rail hybrid vehicles. With battery boost, the diesel engine can be made smaller and lighter and the rail hybrid vehicle can carry less fuel. The decrease of weight of the diesel engine and the cooling system group can be greater than the extra mass of the energy storage.

• Hybrid switchers have a greater success rate compared to other types of rail vehicles. Hybrid switchers operate today in USA, Canada, Great Britain, Sweden,

France, and Japan. The zero emissions switcher with fuel cells will go into service in USA in the fall of 2009.

• Hybrid trains Lirex and Kiha E200 are already deployed in commercial use. East Japan Railway Company is planning to introduce 10 new hybrid diesel trains in 2010. Many hybrid diesel railway cars will be included in high-speed (200km/h) Super Express trains (Great Britain) and will be built in 2012-2013 years.

• The hybrid rail vehicles of the near future will have the heat engine and microturbines, asynchronous drive and energy storage. The hybrid rail vehicles of the more distant future will have the fuel cells, synchronous traction generators and motors with permanent magnets and more effective energy storages.

• The rolling stock today has a high potential for hybridization. Tens of thousands of the freight and switcher locomotives, thousands of diesel trains and vehicles of urban rolling stock wait their hybrid fate. The task is to use this potential.

INDEX

Accumulator's batteries — 33-34, 90-92
Asynchronous traction drive — 42, 45, 47, 49-50, 54, 57, 61,64, 76, 86, 97
Alstom — 8, 13, 33, 44, 48, 82, 85, 88, 90
Ammonia — 28
Asynchronous traction motor — 8, 18, 30, 48, 51, 53-54, 59, 85
Auxiliary system of hybrid rail vehicle — 7, 13-14, 19, 47, 63, 78, 81

Bombardier — 38, 44-45, 52, 54, 56, 74, 85
Braking distance — 89
British High Speed Train (HST) — 78
Brush Traction — 75, 78

Capstone Turbine Corporation — 26
Carbon dioxide — 6, 76
Choppers — 31, 68-69, 73, 80, 85-86
Control system — 12, 62, 81, 97
Converter — 10-11, 14-15, 18-19, 30-32, 38-39, 43-45, 47-48, 50-52, 54, 59, 62-64, 67, 73, 80, 82-83, 85, 93, 96-97

DBAG (Deutshe Bahn AG) — 64
DC electric rail vehicle with on-board energy storage — 56-61
DC link — 8, 14-15, 31, 42, 48, 50-51, 54, 56, 62, 64, 67, 85, 93

Degree of hybridization — 87
Diesel engines — 7-8, 13-15, 25, 28-30, 42-43, 45, 50, 53-54, 63, 65-66, 69-71, 76-747, 79, 85-88, 97

Diesel-generator set — 9, 14, 25, 47, 54, 65, 68-69, 71, 76, 82
Diesel locomotives — 7, 12-13, 19, 25, 48, 50-51, 77, 87, 90, 93, 96

Diesel trains — 6-7, 12, 30, 42, 45, 45, 45, 62, 90, 98
Diode — 8, 30-33, 48, 50
Dual- mode locomotives — 48, 51-52, 54, 72
Dual-mode trains — 42, 47, 79
    AGC regional train — 25, 44-45
    B 81500 — 45-47
    B 82500 — 45-47
Dual-mode tram-trains — 42

Efficiency — 7, 19, 30-32, 34, 56-57, 59, 67-68, 73, 75, 78, 82-83, 93, 95-96, 104, 75,.78, 82-83, 93, 95-96, 104, 116
EMU — 12, 25, 30, 42-43, 58-60, 85-86, 93

Energy flows in the hybrid system
    traction energy flow   9
    braking energy flow   9
    energy recovery flow   10
Energy storage   7-16, 17-20, 22, 28, 31-34, 37-39, 43, 56-60, 66-68, 72-74, 77, 89-80, 85-86, 90, 93, 96-98

EORB (Energy of regenerative braking)   15, 20, 22, 24, 90
eTransformer   55-56
Exhaust emissions   6, 64, 67

FESS (Flywheel Energy StorageSystem)   39
Flywheel   38-39, 51, 65, 67, 67, 74, 79, 85, 90-92, 94-95

Forza Rail Power Module (RPM)
Nuvera Company   82
Freight locomotives   22, 53
Fuel cell   8, 10, 12, 14, 26, 28-29, 79-83, 87-88, 80, 98

General Electric   18, 26, 48, 50, 68, 77
General Motors   12, 50, 76
Grade   15, 18-19, 22, 31-32, 92, 96-97
GTO-thyristor (Gate Turn-Off Thyristor)   31, 50, 82

Hitachi   62, 78-79
Hybrid bus   6, 12, 26, 87
Hybrid Switch Locomotives
    Green Goat (GG 20B)   68-71, 78-76, 86-87, 91
    Green Kid (GK10B)   71, 86-87
    Road Switcher RP20BH   70-72, 87
Hybrid rail vehicles in the operation
    First in the world
    hybrid railcar (Japan)   62, 64
    Regional innovative train LIREX
    (Germany)   25, 30, 44-45, 47, 64-67, 72, 85, 90-93, 95-97
    Railcars PPM 50 and PPM 80 (England)   67-68, 90
Hybrid rail vehicles with fuel cells (prototypes)
    Hybrid fuel cell tram
    (FULLTRAM project)   79-80
    World's First Fuel Cell Hybrid
    Railcar (East Japan Railway)   80-81
    Fuel cell train project of Japan
    Railway Technical Research
    Institute (RTRI)   81-82

Fuel cell-battery Hybrid Switcher 82-83, 97
Hybrid rail vehicles with heat engine
(prototypes)
    Hybrid tram with microturbine and
    flywheel energy storage(project) 73-74
    Hybrid diesel train with
    ultracapacitor's energy storage
    (Mitrac Energy Saver,
    Bombardier Germany) 74
    English Green Goat switcher 75
    Swedish Green Goat switcher 76
    Hybrid modernized switcher
    locomotives of Alstom Transport
    (France) 76
    Turbine-electric switcher locomotive
    (Russia) 77
    GE hybrid diesel locomotive (USA) 77-78
Hydrogen 11, 27-28, 35, 79. 81-83

IGBT (Insulated gate bipolar transistors) 18, 30-31, 39, 43-45, 50, 52-53, 56, 64, 73-74, 78, 85, 93, 97

Japan Rail East 62, 64, 80-81

Kinetic energy of the vehicle 11-12, 15-16, 18, 20, 22, 38, 8-58, 67
Kyoto Commitment 6

Lead Acid Battery 33-35, 68-69, 87
Lithium Ion Battery 36, 58-59, 63, 78, 85
Lithium Ion Polymer Battery 34, 37
Lubricants 18, 88

Magnetic bearing 37
Main generators 29, 42, 53, 85, 97
Methane 6, 25
Methanol 28, 82
Microprocessor 68, 76, 87
Microturbine 25-26, 80, 98
Mitrac Energy Saver 56-57, 74, 85, 91
Modifiers 18
MOSFET (low power consumption and
high switching transistor) 31
MTU 25, 49-50, 78

(NeFeB) neodymium boron iron (NeFeB) 30
Nickel Metal Hydride (Ni MH) battery 31, 35, 57, 77, 91-92, 94-95

Parry People Movers Ltd.   67
PEM (Proton exchange membrane)
fuel cell   26-27, 28, 81-83, 88
Permanent magnet generator   26, 29
PORB (Power of regenerative braking)   20-24, 90
Prime movers and motors   25
PWM (pulse width modulated) inverter   8, 31, 33, 39, 43, 48, 50, 59, 64, 76

Rail Power Technology   68
Rechargeable metal hydrides   28
Rectifier   31, 46-47, 53-54, 65
Regenerative braking   7, 15, 18-20, 22-23, 50, 56, 58-60, 62, 64, 77, 88. 90, 93, 97
Rolling resistance   12-13, 15, 18-20, 97
Rolling stock   11, 18, 20, 32, 60, 71, 97
RTRI (Japan Railway Technical Research Institute)   58-60, 81-82, 68-70, 99-100

Siemens   42, 49-51
Slope   18-19, 62, 96-97
Specific energy of different storage device families   34-35, 38-39, 90
Specific power of different storage devices families   34, 90
Switcher locomotives   14, 22, 76, 87, 98
Synchronous generator   29, 39, 45

Tokyo Car Corporation   62
Toyota Prius   12, 86, 91
Traction motors   8, 12, 14, 16, 18, 20, 30-32, 42, 48, 50-53, 56, 59, 63-65, 68-69, 71, 73, 75, 77, 82, 85, 87, 97
Tram-trains   12, 14, 21, 42-43
    Avanto (S70) (Siemens)   42-43
    Combino (Siemens)   42-43
    Regio Citadis   42-45
Transformer   12, 14, 21, 42-43, 46-48, 54, 75

Ultracapacitors   7, 31-32, 37-38, 56, 74, 77, 79-80, 82, 90-92, 94-97
UN definition of hybrid rail vehicles   10

Valve Regulated Lead Acid
(VRLA) batteries   34, 90-91

ZEBRA (Sodium/Nickel – Chloride)
Battery   34-36, 91

9. LIST OF TABLES

2.1. Main characteristics modern accumulator's batteries

3.1. European tram-trains - main characteristics

3.2. Main characteristics of the modern diesel trains.

3.3. Main characteristics of the modern diesel locomotives with asynchronous traction motors

3.4. Main characteristics of the modern dual-mode locomotives

5.1. Main technical characteristics of the Lirex train

5.2. Technical data of the Lirex flywheel

5.3. Main parameters of the hybrid locomotives GK10B, GG20B, and RP20BH

6.1. Main parameters of the hybrid locomotive British Green Goat

7.1. The main characteristics of the hybrid diesel electric multiple units (HDEMU)

7.2. Comparison of power ratings for both heat-engine and electric motors of hybrid switchers, trains, buses, and autos.

7.3. Value q variation from speed and braking distance of hybrid switcher.

7.4. Comparison table of modern energy storage systems

7.5. Brake energy and brake power for different rail vehicle types

7.6. The actual mass of energy storage systems for five different types of hybrid rail vehicles

7.7. The mass of energy storage systems with combined use of battery packs and ultracapacitors

10. THE LIST OF THE FIGURES

1.1. Main circuits of the hybrid rail vehicle with heat engine and AC drive
1.2. Main circuits of the hybrid rail vehicle with fuel cells and AC drive
1.3. Energy flow- traction mode
1.4. Energy flow- regenerative mode
1.5. Energy flow- recovery mode
1.6. The rolling stock with the energy storage system.
1.7. First variant of auxiliary system circuit
1.8. Second variant of auxiliary system circuit
1.9. Plots of coefficient of decrease of energy consumption as a function of number of stops and the energy storage system's efficiency
1.10. Plots of coefficient of decrease of energy consumption as a function of number of stops and the value "q".
1.11. EORB for trams, tram-trains, railcars.
1.12. PORB for 50t weight vehicle.
1.13. PORB for 100t weight vehicle.
1.14. PORB for 250t weight vehicle.
1.15. EORB of the switcher hybrid locomotives.
1.16. PORB for the switcher hybrid locomotives
1.17. EORB per 1 km by route. Hybrid locomotive run with the freight train.
1.18. PORB per 1 km by route. Hybrid freight locomotive with the train (grade is 0.5%)
1.19. PORB per 1 kmby route. Hybrid freight locomotive with the train (grade is 2.0%)
2.1. Diagram of diesel-generator power source with schema of the rectifier
2.2. Block diagram of energy storage with ultracapacitors
2.3. Block diagram of asynchronous traction drive with PWM inverter
2.4. Block diagram of a flywheel energy storage system
3.1. Main power circuits - dual-mode variant - Regio Citadis dual-mode tram.
3.2. Block diagram. Power circuits - dual-mode train AGS in diesel mode
3.3. Block diagram. Power circuits of the dual-mode and dual system train B82500 (France)
3.4. Block diagram. Power circuits - dual-mode train - Lirex (Germany)
3.5. Block diagram. Main circuit - SD 70 MAC locomotive
3.6. Main circuit diagram of the locomotives 2016 and 475000.
3.7. The main circuit diagram of the dual- mode locomotive model 38 (South Africa)
3.8. Main circuit diagram of the dual- mode locomotive model DM30AC (USA).
3.9. Main circuit diagram (half) of the dual-mode locomotive model CC 3600 (Spain)
4.1. Block diagram of the energy system for the energy recycling tram and EMU with Li-Ion battery energy storage (Japan)
4.2. Block diagram of the PPM 35 vehicle
5.1. The diagram of the series hybrid system of the first in the world hybrid diesel train (Japan)

5.2. Diagram. The power circuits of the Lirex train in diesel mode.

5.3. The block diagram of the PPM 50 hybrid vehicle

5.4. Bloc diagram of Green Goat power circuits

6.1. Block diagram of ULEV-TAP hybrid tram.

6.2. Block diagram of England's hybrid High Speed Train

6.3. Block diagram of the fuel cell tram.

6.4. Block diagram of the fuel cell train (Japan)

6.5. Block diagram of the RTRI hybrid fuel cell train (Japan)

6.6. Block diagram of the fuel cell hybrid switcher (USA)

7.1. Specific energy of different energy storage systems

7.2. Specific power of different energy storage systems